Mathematics

Essential Skills

8

OXFORD
UNIVERSITY PRESS

Oxford University Press is a department of the University of Oxford. It furthers the University's objective of excellence in research, scholarship, and education by publishing worldwide. Oxford is a registered trademark of Oxford University Press in the UK and in certain other countries.

Published in Australia by
Oxford University Press
8/737 Bourke Street, Docklands, Victoria 3008, Australia

First published 2011
Reprinted 2019(D)

ISBN 978 0 19 557601 6

Edited by Emma Short
Typeset by Palmer Higgs
Printed and bound in Australia by Ligare Book Printers Pty Ltd

PAPUA NEW GUINEA

Mathematics

Essential Skills

8

Pat Lilburn

Contents

For Students

The *Oxford Essential Skills Book* has been written to support the *Oxford Grade 8 Mathematics Student Books A* and *B*.

The focus of this book is on providing practice examples to revise and support the concepts learned in the two Student Books.

The book is set out under the five Strands outlined in the *Upper Primary Mathematics Syllabus* and *Teacher Guide*:

- Number and Application
- Space and Shape
- Measurement
- Chance and Data
- Patterns and Algebra

Within each Strand, you will find many practice examples that relate directly to the Learning Outcomes for that Strand. These Learning Outcomes are specified in the *Upper Primary Mathematics Syllabus*.

The subheading in each Strand shows you what the examples will help you practise.

8.1.7 Apply directed numbers in problem solving

The Help Box contains worked examples to show you how to set out and complete the examples in each Strand. The Remember Box contains information that will help you complete the examples.

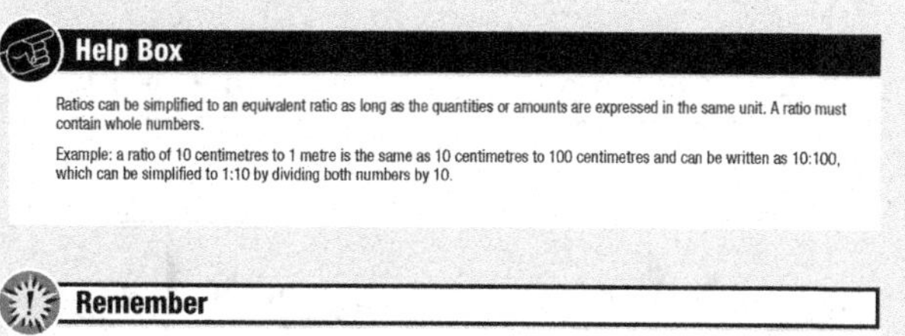
Help Box

Ratios can be simplified to an equivalent ratio as long as the quantities or amounts are expressed in the same unit. A ratio must contain whole numbers.

Example: a ratio of 10 centimetres to 1 metre is the same as 10 centimetres to 100 centimetres and can be written as 10:100, which can be simplified to 1:10 by dividing both numbers by 10.

Remember

When multiplying decimals by multiples of 10, first multiply by 10 and then multiply by the number of tens.

At the end of each Strand, there is an Assessment section that covers all the examples from that Strand. If you have difficulty with any test questions, go back to the relevant page in the Strand and look at the worked examples before doing the test again.

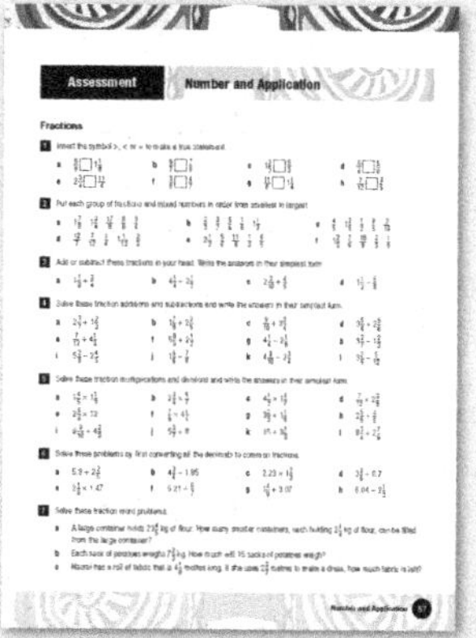
Assessment Number and Application

Fractions

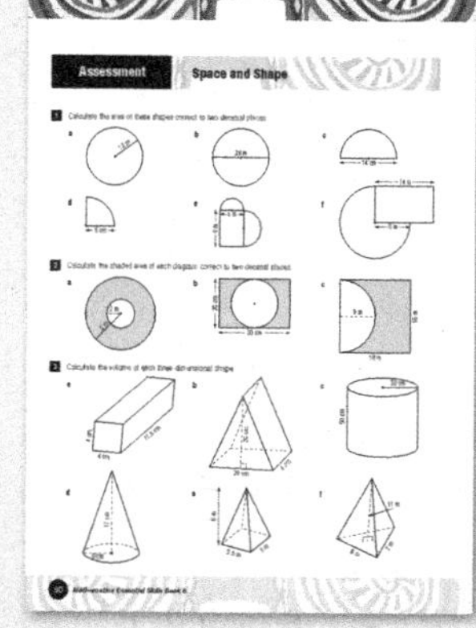
Assessment Space and Shape

There is an Answers section and a section for Important Facts at the back of the book. The Answers section contains answers to all practice examples and test questions while the Important Facts section contains some of the facts that you will need to refer to throughout the year.

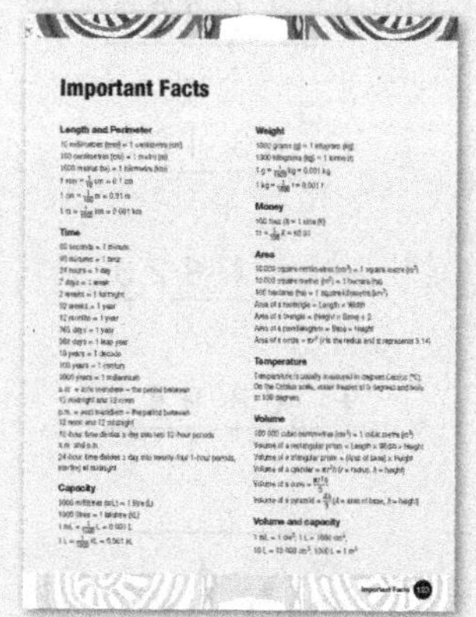
Important Facts

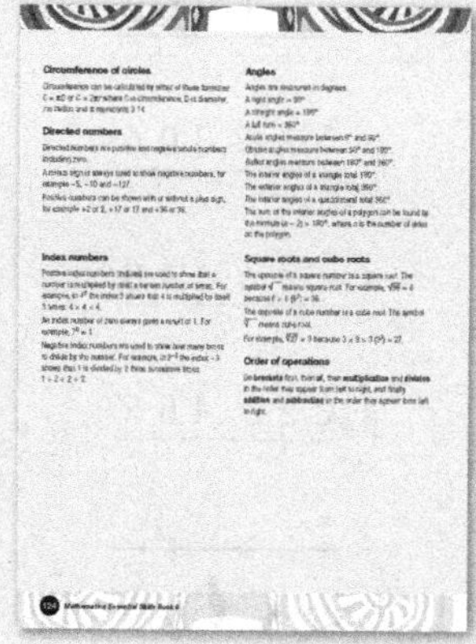

Strand | Number and Application

Revise addition with whole numbers

Copy and add at each stage of the addition columns until you reach the number at the bottom. Did you get the answer shown for each column?

1	2	3	4
43 245	250 825	126 193	353 226
+ 54 636	+ 38 851	+ 43 567	+ 7 946
+ 20 457	+ 62 339	+ 35 931	+ 93 448
+ 8 927	+ 56 928	+ 87 046	+ 224 032
+ 33 084	+ 7 504	+ 211 834	+ 47 614
+ 216 580	+ 96 373	+ 8 968	+ 11 693
+ 69 933	+ 32 864	+ 60 535	+ 44 121
+ 59 284	+ 3 091	+ 42 308	+ 68 917
+ 88 439	+ 27 499	+ 15 639	+ 154 768
+ 9 968	+ 44 357	+ 443 822	+ 39 866
+ 78 604	+ 83 105	+ 24 377	+ 9 879
+ 97 116	+ 4 823	+ 84 916	+ 52 606
+ 328 455	+ 68 949	+ 4 398	+ 94 825
	777 508	1 189 534	1 202 941

Revise subtraction with whole numbers

1 Copy and complete the chart by subtracting the smaller numbers from the larger numbers.

	–	58 376	109 216	39 857	345 862	83 958
a	84 558					
b	216 033					
c	177 525					
d	892 481					
e	67 304					
f	58 547					
g	308 164					
h	148 293					

2 Use subtraction to find the missing numbers.

a
```
   76 381
 – □□ □□□
   45 297
```

b
```
  254 165
– □□□ □□□
  178 449
```

c
```
  407 362
– □□□ □□□
  289 556
```

d
```
  128 003
–  □□ □□□
   68 572
```

e
```
  □□□ □□□
+ 241 667
  532 818
```

f
```
  □□□ □□□
+  75 824
  232 969
```

g
```
  □□□ □□□
+ 109 475
  316 751
```

h
```
  □□□ □□□
+  58 084
  195 336
```

i
```
   66 473
+ □□□ □□□
  354 089
```

j
```
  685 019
+ □□□ □□□
  817 433
```

k
```
  428 360
+ □□□ □□□
  755 821
```

l
```
  376 508
+ □□□ □□□
  639 427
```

Revise multiplication with whole numbers

1 Solve these multiplications.

a 28 453 × 8	**b** 63 772 × 6	**c** 45 908 × 9	**d** 35 627 × 4
e 54 748 × 7	**f** 223 196 × 8	**g** 185 334 × 5	**h** 782 035 × 3
i 72 766 × 19	**j** 42 894 × 22	**k** 28 496 × 35	**l** 59 057 × 28
m 51 534 × 24	**n** 19 781 × 18	**o** 67 998 × 34	**p** 35 483 × 43
q 95 257 × 48	**r** 88 209 × 36	**s** 26 556 × 25	**t** 47 618 × 39
u 63 447 × 16	**v** 31 889 × 54	**w** 74 066 × 27	**x** 92 352 × 33

2 Use the answers to:

a subtract the smallest product from the largest product

b find the total of the five largest products

c find the difference between the two smallest products

Revise division with whole numbers

1 Use division to find the missing numbers.

a □□□□ × 8 68432	**b** □□□□ × 5 39725	**c** □□□□ × 7 48139	**d** □□□□ × 3 25428
e □□ □□□ × 4 69 584	**f** □□ □□□ × 9 253 467	**g** □□ □□□ × 6 256 698	**h** □□ □□□ × 8 505 744
i □□□□ × 15 37560	**j** □□□□ × 23 38962	**k** □□□□ × 18 74430	**l** □□□□ × 35 121660
m □□□□ × 27 177849	**n** □□□□ × 32 91456	**o** □□□□ × 45 184320	**p** □□□□ × 29 157905
q □□□□ × 25 44475	**r** □□□□ × 17 64889	**s** □□□□ × 36 92376	**t** □□□□ × 22 160248

2 Solve these divisions and write any remainders as fractions reduced to their lowest terms.

a 8)7605	**b** 5)8117	**c** 7)5966	**d** 6)9254
e 4)25 088	**f** 9)34 712	**g** 8)15 677	**h** 5)34 098
i 7)512 674	**j** 3)827 928	**k** 9)416 079	**l** 8)230 646
m 12)33 154	**n** 25)57 836	**o** 16)26 551	**p** 32)65 458
q 51)62 595	**r** 15)49 066	**s** 28)72 173	**t** 19)95 302

Use the four processes to revise word problems

1 In 2004, there were 33 255 visitors from Australia to Papua New Guinea. This number increased to 54 098 in 2007. How many more visitors from Australia were there in 2007 than in 2004?

2 A plane travels 14 467 kilometres on a one-way flight between Port Moresby and London. How many kilometres will the plane travel if it completes the following numbers of one-way flights?

a 3 **b** 7 **c** 4 **d** 9 **e** 6 **f** 8

3 Over a two-week period, a car salesman sold cars to the value of K10 466, K12 021, K9875, K6337 and K11 874.

a What was the total value of the cars sold?

b What was the mean price of a car sold?

4 In seven days, an athlete walked 164 192 steps.

a What was the mean number of steps walked by the athlete each day?

b Use this mean to calculate the number of steps the athlete might walk in 18 days.

5 If a family pays K3468 per month to rent a house, how much will they pay for the following numbers of months?

a 6 **b** 4 **c** 12 **d** 9 **e** 18 **f** 15

6 At the port, two containers are loaded on to a ship. One container weighs 17 784 kilograms and the other weighs 25 433 kilograms.

a What is the difference in weight between the two containers?

b What is the total weight of the two containers?

7 The population of six countries is shown in this chart.

Country	Population	Country	Population
Cyprus	801 851	Malta	416 333
Solomon Islands	530 669	Iceland	318 236
Samoa	187 032	Montenegro	641 966

a How many more people live in Cyprus than in Samoa?

b What is the combined population of Malta and Montenegro?

c How many more people live in the Solomon Islands than in Iceland?

d If the population of Malta tripled, what would it be?

e If the population of the Solomon Islands doubled, what would it be?

8 A syndicate of 16 people equally shared a lottery win of K194 012. How much money did each person receive?

9 A holiday package costs K1735 per adult and K1089 per child. If a family of two adults and three children purchase this package, how much will it cost in total?

Revise and practise number facts

Write answers for each multiplication and division. Try to do them as quickly as you can.

1	2	3	4	5
5×9	$21 \div 3$	3×3	$60 \div 6$	3×7
2×2	$42 \div 7$	9×8	$96 \div 12$	12×6
6×7	$63 \div 7$	7×4	$25 \div 5$	8×9
4×8	$72 \div 9$	10×12	$36 \div 4$	7×5
3×7	$9 \div 9$	12×6	$24 \div 6$	6×9
10×4	$20 \div 4$	6×9	$40 \div 5$	8×7
8×5	$32 \div 4$	3×5	$110 \div 11$	4×2
3×4	$36 \div 9$	4×6	$10 \div 5$	0×8
6×3	$56 \div 8$	7×7	$18 \div 2$	5×6
8×9	$27 \div 9$	6×3	$20 \div 10$	4×8
9×11	$45 \div 5$	2×9	$4 \div 4$	9×7
12×12	$18 \div 6$	10×10	$16 \div 2$	3×8
6×6	$12 \div 2$	12×1	$27 \div 9$	12×4
4×0	$16 \div 4$	6×6	$72 \div 8$	5×11
3×8	$25 \div 5$	1×8	$80 \div 8$	7×8
9×6	$70 \div 7$	0×7	$144 \div 12$	5×5
5×5	$99 \div 9$	11×11	$54 \div 6$	4×7
11×8	$8 \div 4$	3×8	$27 \div 3$	9×3
3×3	$24 \div 2$	4×6	$9 \div 3$	7×5
9×0	$36 \div 3$	5×9	$15 \div 5$	2×6
1×5	$48 \div 6$	8×12	$21 \div 7$	11×6
2×11	$54 \div 6$	4×4	$45 \div 5$	8×8
4×4	$30 \div 5$	2×8	$60 \div 12$	8×4
8×8	$35 \div 7$	7×8	$77 \div 11$	3×0
6×5	$50 \div 10$	3×6	$28 \div 7$	12×5
4×9	$121 \div 11$	6×9	$30 \div 6$	6×6
6×8	$96 \div 8$	4×7	$81 \div 9$	9×8
8×2	$108 \div 12$	6×8	$90 \div 10$	7×10
7×7	$24 \div 8$	0×4	$121 \div 11$	3×11
10×4	$45 \div 9$	3×10	$96 \div 8$	7×6

6	7	8	9	10
6 × 9	30 ÷ 5	6 × 9	48 ÷ 6	4 × 8
5 × 5	45 ÷ 9	3 × 3	18 ÷ 6	9 × 3
2 × 9	66 ÷ 6	2 × 8	36 ÷ 9	2 × 1
8 × 3	42 ÷ 7	11 × 6	40 ÷ 4	3 × 4
7 × 8	63 ÷ 9	0 × 5	44 ÷ 11	5 × 6
9 × 8	110 ÷ 11	1 × 2	56 ÷ 7	7 × 8
7 × 12	56 ÷ 7	3 × 8	30 ÷ 3	3 × 6
6 × 6	12 ÷ 12	10 × 11	96 ÷ 8	7 × 7
4 × 9	5 ÷ 1	12 × 3	64 ÷ 8	9 × 8
12 × 9	9 ÷ 3	4 × 9	42 ÷ 7	10 × 7
4 × 5	25 ÷ 5	5 × 7	15 ÷ 5	11 × 1
7 × 6	60 ÷ 12	6 × 4	10 ÷ 2	9 × 9
5 × 9	96 ÷ 12	7 × 7	54 ÷ 6	3 × 8
6 × 11	54 ÷ 6	9 × 11	63 ÷ 7	5 × 9
12 × 8	18 ÷ 9	12 × 6	21 ÷ 3	2 × 10
3 × 6	27 ÷ 3	4 × 5	66 ÷ 6	6 × 6
7 × 3	24 ÷ 8	1 × 9	72 ÷ 9	7 × 8
5 × 12	36 ÷ 6	2 × 6	72 ÷ 6	8 × 5
1 × 7	64 ÷ 8	12 × 10	84 ÷ 7	9 × 11
4 × 8	42 ÷ 6	2 × 8	99 ÷ 9	4 × 4
11 × 12	48 ÷ 8	3 × 6	36 ÷ 3	6 × 8
6 × 4	16 ÷ 4	8 × 4	45 ÷ 5	11 × 5
3 × 8	12 ÷ 3	8 × 8	88 ÷ 11	6 × 3
7 × 7	30 ÷ 3	7 × 6	120 ÷ 12	4 × 9
5 × 7	56 ÷ 8	5 × 7	108 ÷ 12	8 × 3
9 × 6	88 ÷ 11	3 × 9	36 ÷ 4	11 × 12
8 × 11	72 ÷ 9	9 × 9	81 ÷ 9	10 × 0
7 × 4	90 ÷ 10	10 × 1	12 ÷ 6	5 × 6
8 × 9	108 ÷ 12	5 × 9	24 ÷ 2	7 × 5
6 × 8	21 ÷ 7	12 × 12	132 ÷ 11	8 × 7

8.1.1 Apply fractions in problem solving

Compare and order fractions

1 Copy these pairs of fractions or mixed numbers and insert the symbol > or < to make a true statement.

a $\frac{1}{4} \square \frac{3}{8}$ b $\frac{8}{9} \square \frac{2}{3}$ c $1\frac{1}{2} \square \frac{5}{2}$ d $\frac{2}{5} \square \frac{3}{10}$

e $2\frac{1}{3} \square \frac{5}{3}$ f $\frac{5}{6} \square \frac{3}{9}$ g $\frac{8}{7} \square \frac{7}{8}$ h $\frac{3}{4} \square \frac{4}{3}$

i $\frac{7}{10} \square \frac{3}{5}$ j $2\frac{3}{4} \square \frac{9}{4}$ k $\frac{4}{5} \square \frac{9}{10}$ l $\frac{1}{6} \square \frac{1}{3}$

m $\frac{2}{9} \square \frac{1}{2}$ n $\frac{7}{11} \square \frac{7}{10}$ o $\frac{5}{6} \square \frac{7}{8}$ p $1\frac{4}{5} \square 1\frac{2}{3}$

2 Which fraction is larger? Convert one fraction in each pair so that you can compare them to find the larger one.

a $\frac{13}{8}$ or $1\frac{3}{4}$ b $1\frac{2}{3}$ or $\frac{13}{9}$ c $\frac{22}{7}$ or $3\frac{1}{5}$ d $1\frac{3}{4}$ or $\frac{15}{8}$ e $\frac{30}{9}$ or $3\frac{2}{5}$ f $2\frac{5}{8}$ or $\frac{13}{5}$

g $\frac{23}{9}$ or $2\frac{4}{7}$ h $2\frac{2}{3}$ or $\frac{14}{5}$ i $\frac{26}{5}$ or $5\frac{1}{4}$ j $1\frac{5}{11}$ or $\frac{11}{7}$ k $\frac{29}{6}$ or $4\frac{9}{10}$ l $1\frac{1}{6}$ or $\frac{6}{5}$

m $\frac{27}{8}$ or $3\frac{5}{7}$ n $4\frac{3}{10}$ or $\frac{13}{3}$ o $\frac{25}{11}$ or $2\frac{4}{9}$ p $3\frac{7}{8}$ or $\frac{32}{9}$ q $\frac{20}{6}$ or $3\frac{1}{7}$ r $5\frac{2}{3}$ or $\frac{23}{4}$

3 Write a fraction that comes between each pair of fractions. For example, a fraction between $\frac{1}{3}$ and $\frac{7}{8}$ could be $\frac{1}{2}$.

a $\frac{1}{3}$ and $\frac{7}{8}$ b $\frac{1}{2}$ and $\frac{3}{4}$ c $\frac{3}{10}$ and $\frac{4}{5}$ d $\frac{2}{5}$ and $\frac{2}{3}$ e $\frac{3}{8}$ and $\frac{7}{10}$ f $\frac{3}{5}$ and $\frac{3}{4}$

g $\frac{1}{8}$ and $\frac{1}{2}$ h $\frac{2}{7}$ and $\frac{5}{8}$ i $\frac{1}{4}$ and $\frac{1}{2}$ j $\frac{1}{5}$ and $\frac{4}{7}$ k $\frac{2}{3}$ and $\frac{8}{9}$ l $\frac{1}{6}$ and $\frac{1}{2}$

m $\frac{2}{9}$ and $\frac{5}{8}$ n $\frac{1}{2}$ and $\frac{5}{6}$ o $\frac{2}{3}$ and $\frac{9}{10}$ p $\frac{5}{12}$ and $\frac{3}{4}$ q $\frac{1}{7}$ and $\frac{3}{8}$ r $\frac{1}{4}$ and $\frac{5}{12}$

4 Put each group of fractions and mixed numbers in order from smallest to largest.

a $\frac{3}{5}$ $\frac{2}{3}$ $\frac{1}{4}$ $\frac{5}{3}$ $\frac{1}{10}$ b $2\frac{3}{4}$ $\frac{7}{12}$ $\frac{3}{8}$ $\frac{1}{3}$ $\frac{7}{4}$ c $\frac{10}{5}$ $\frac{3}{4}$ $1\frac{7}{8}$ $\frac{5}{6}$ $\frac{5}{8}$

d $\frac{4}{5}$ $\frac{9}{9}$ $\frac{2}{3}$ $1\frac{2}{7}$ $\frac{5}{7}$ e $1\frac{1}{3}$ $\frac{8}{9}$ $\frac{1}{2}$ $\frac{1}{8}$ $\frac{3}{7}$ f $\frac{9}{10}$ $\frac{7}{11}$ $\frac{8}{7}$ $1\frac{1}{8}$ $\frac{4}{9}$

g $\frac{10}{7}$ $\frac{7}{10}$ $\frac{11}{12}$ $\frac{6}{3}$ $\frac{4}{7}$ h $\frac{15}{5}$ $2\frac{1}{2}$ $\frac{13}{8}$ $1\frac{3}{4}$ $\frac{4}{3}$ i $\frac{2}{7}$ $\frac{2}{9}$ $\frac{2}{3}$ $\frac{2}{1}$ $\frac{1}{2}$

Add fractions

1 Add these fractions in your head and write the answers reduced to their simplest form.

a $2\frac{1}{4} + 3\frac{3}{4}$ b $\frac{7}{8} + 1\frac{1}{2}$ c $2\frac{3}{4} + \frac{4}{8}$ d $\frac{5}{8} + \frac{1}{4}$ e $3\frac{1}{2} + \frac{5}{8}$ f $1\frac{1}{8} + 2\frac{1}{4}$

g $4\frac{7}{8} + 1\frac{1}{2}$ h $3\frac{3}{4} + 1\frac{3}{8}$ i $5\frac{1}{2} + 6\frac{3}{4}$ j $3\frac{1}{3} + 2\frac{2}{3}$ k $1\frac{2}{3} + 1\frac{2}{3}$ l $4\frac{2}{3} + 3\frac{1}{6}$

m $2\frac{1}{5} + 4\frac{3}{10}$ n $1\frac{4}{5} + 3\frac{4}{5}$ o $2\frac{7}{10} + \frac{3}{5}$ p $\frac{5}{6} + 2\frac{1}{3}$ q $4\frac{1}{2} + 1\frac{1}{6}$ r $5\frac{2}{3} + \frac{5}{9}$

2 Copy and complete the chart by calculating the total of each pair of fractions.

	+	$2\frac{5}{8}$	$1\frac{4}{5}$	$4\frac{1}{9}$	$2\frac{5}{7}$	$3\frac{2}{11}$
a	$3\frac{4}{7}$					
b	$1\frac{7}{12}$					
c	$2\frac{3}{5}$					
d	$4\frac{3}{8}$					
e	$1\frac{5}{6}$					
f	$2\frac{7}{9}$					
g	$3\frac{7}{10}$					
h	$4\frac{2}{3}$					

3 Which total is the largest? Which total is the smallest?

4 Add these fractions and mixed numbers. Write all answers in their simplest form.

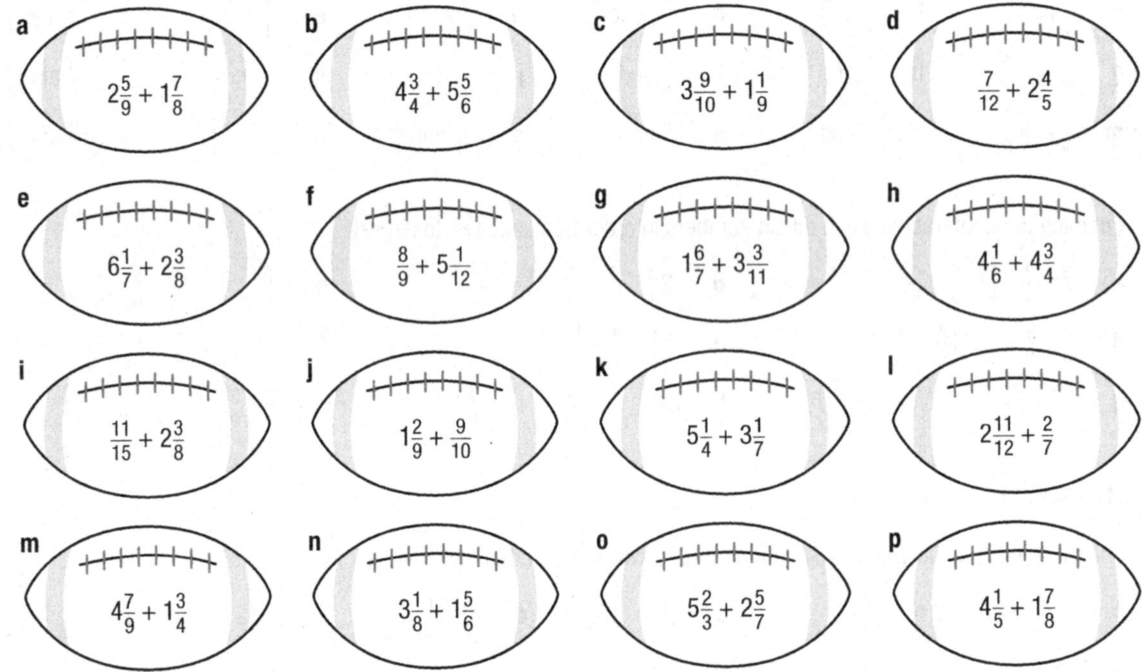

How many goals did you shoot?

Subtract fractions

1 Subtract these fractions in your head and write the answers reduced to their simplest form.

a $3\frac{1}{2} - \frac{5}{8}$	**b** $4\frac{1}{4} - 2\frac{1}{2}$	**c** $5\frac{3}{8} - 2\frac{3}{4}$	**d** $2\frac{1}{8} - 1\frac{1}{4}$
e $3\frac{3}{4} - \frac{7}{8}$	**f** $1\frac{1}{2} - \frac{3}{8}$	**g** $4\frac{1}{4} - \frac{5}{8}$	**h** $6\frac{1}{2} - 2\frac{1}{6}$
i $3\frac{5}{6} - 1\frac{1}{2}$	**j** $3\frac{2}{3} - 1\frac{5}{6}$	**k** $4\frac{1}{6} - 1\frac{2}{3}$	**l** $3\frac{2}{3} - 1\frac{5}{9}$
m $5\frac{4}{9} - 2\frac{1}{3}$	**n** $3\frac{1}{9} - \frac{2}{3}$	**o** $4\frac{3}{10} - 2\frac{2}{5}$	**p** $3\frac{1}{5} - \frac{9}{10}$
q $2\frac{1}{10} - 1\frac{1}{2}$	**r** $6\frac{1}{2} - 1\frac{7}{10}$	**s** $4\frac{7}{10} - 2\frac{4}{5}$	**t** $3\frac{2}{9} - 1\frac{1}{3}$
u $\frac{11}{12} - \frac{3}{4}$	**v** $2\frac{1}{12} - 1\frac{1}{6}$	**w** $5\frac{1}{3} - 2\frac{5}{12}$	**x** $4\frac{1}{4} - \frac{7}{12}$

2 Complete the missing fractions in each grid. Work out the answer to the first subtraction and place the answer in the space below so that you can calculate the answer to the second subtraction.

a

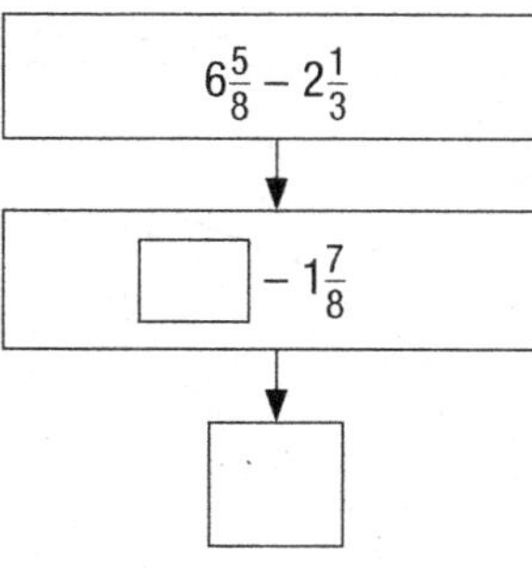

b

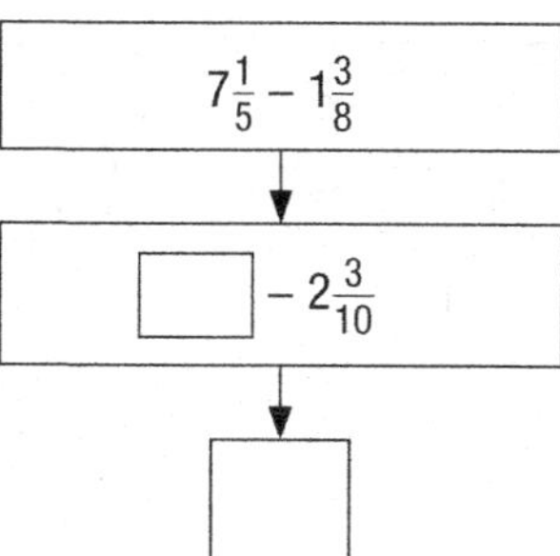

c

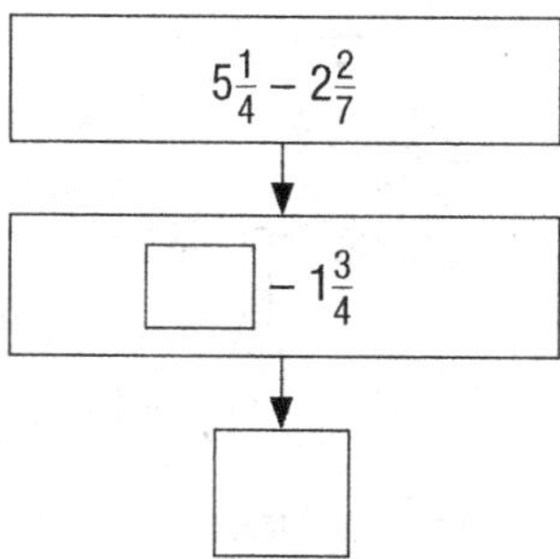

d

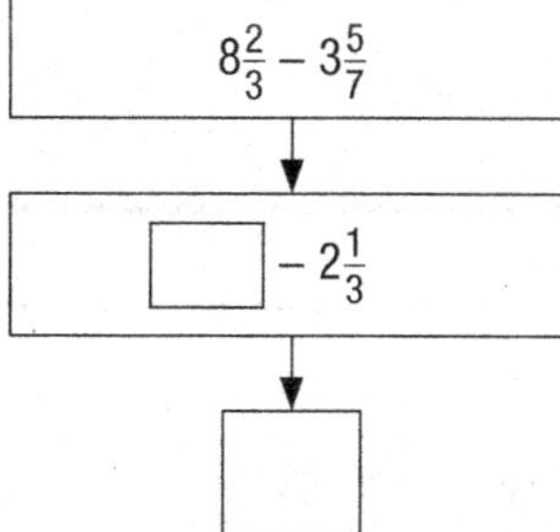

e

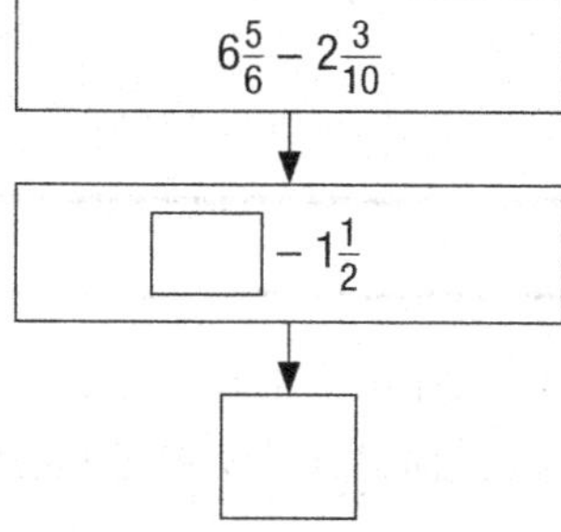

f

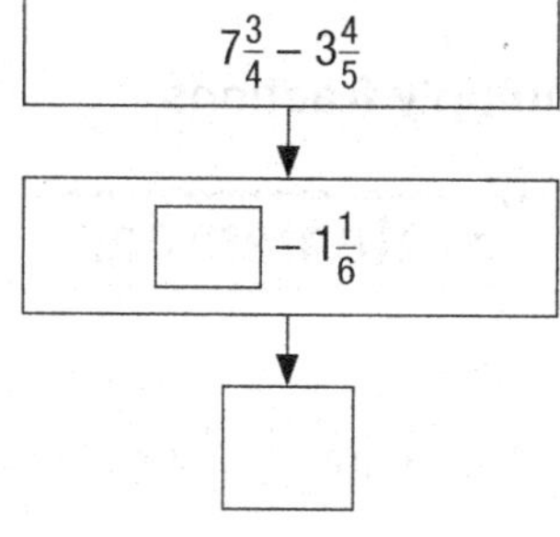

g

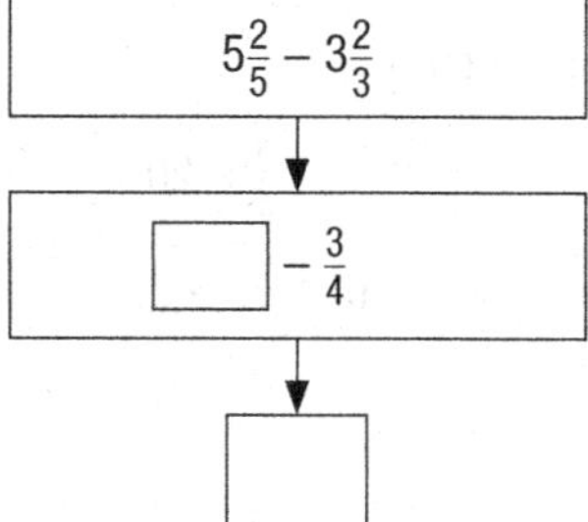

h

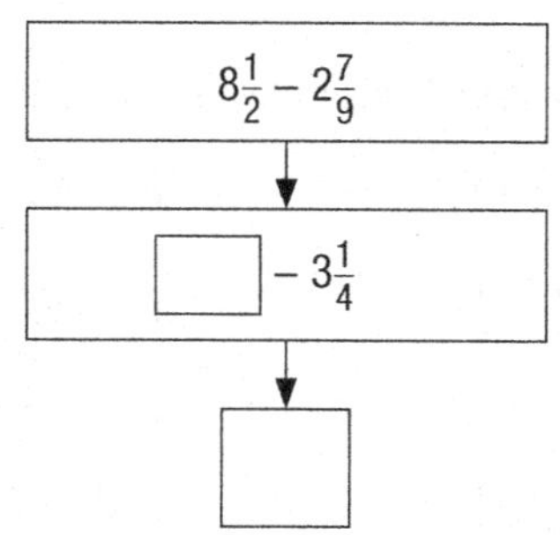

i

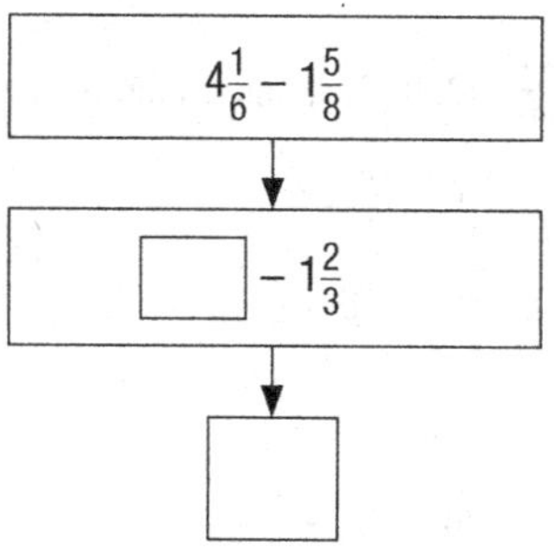

3 Subtract these fractions and mixed numbers. Write all answers in their simplest form.

a
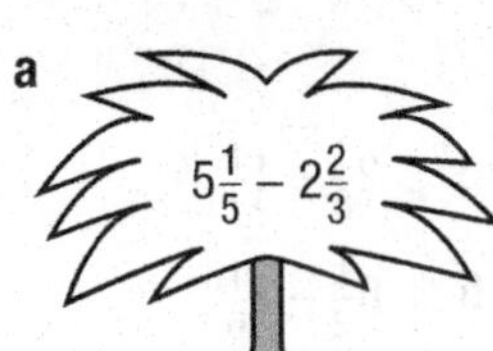

b

c

d

e

f
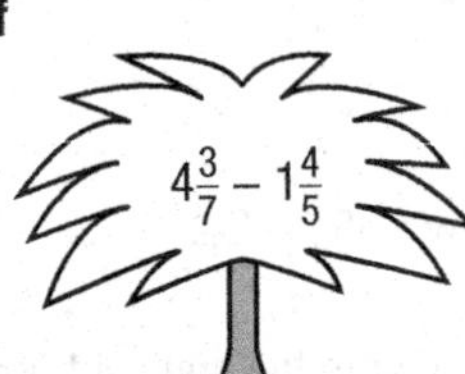

g

h

i

$6\frac{2}{3} - 2\frac{7}{10}$

j

k

l

How many trees did you climb?

Multiply fractions

Remember

When mentally multiplying fractions, the fractions can be multiplied in any order and the '×' symbol means the same as 'of'.

For example, $7\frac{1}{2} \times \frac{1}{3}$ is the same as $\frac{1}{3} \times 7\frac{1}{2}$ and means $\frac{1}{3}$ of $7\frac{1}{2}$. This can be calculated mentally as $\frac{1}{3}$ of $7\frac{1}{2} = 2\frac{1}{2}$.

1 Multiply these fractions in your head and write the answers reduced to their simplest form.

a	$\frac{1}{2} \times 6$	**b**	$\frac{1}{4} \times 12$	**c**	$\frac{1}{3} \times 9$	**d**	$\frac{1}{5} \times 20$
e	$16 \times \frac{1}{4}$	**f**	$30 \times \frac{1}{3}$	**g**	$18 \times \frac{1}{6}$	**h**	$24 \times \frac{1}{8}$
i	$\frac{1}{2} \times 2\frac{1}{2}$	**j**	$\frac{1}{2} \times 8\frac{2}{3}$	**k**	$\frac{1}{2} \times 4\frac{2}{5}$	**l**	$\frac{1}{2} \times 2\frac{6}{7}$
m	$3\frac{3}{7} \times \frac{1}{3}$	**n**	$6\frac{3}{4} \times \frac{1}{3}$	**o**	$10\frac{5}{8} \times \frac{1}{5}$	**p**	$8\frac{4}{7} \times \frac{1}{4}$

2 Copy and complete the chart by calculating the product of each pair of fractions.

	×	$1\frac{7}{8}$	$2\frac{3}{4}$	$5\frac{4}{5}$	$3\frac{2}{3}$	$1\frac{3}{7}$
a	$2\frac{1}{3}$					
b	$1\frac{3}{4}$					
c	$8\frac{1}{6}$					
d	$4\frac{7}{10}$					
e	$2\frac{3}{8}$					
f	$1\frac{2}{5}$					
g	$5\frac{1}{2}$					
h	$3\frac{5}{7}$					

3 Multiply these fractions. Start at the bottom of each ladder.

a

$\frac{4}{5} \times 1\frac{3}{8}$

$1\frac{2}{3} \times 3\frac{3}{4}$

$2\frac{1}{5} \times 2\frac{5}{7}$

$1\frac{3}{10} \times \frac{5}{8}$

$4\frac{1}{4} \times 6\frac{3}{5}$

$1\frac{2}{11} \times \frac{3}{4}$

$5\frac{1}{2} \times 3\frac{5}{6}$

$14 \times 3\frac{5}{8}$

$2\frac{5}{11} \times 3\frac{3}{7}$

b

$2\frac{3}{4} \times \frac{5}{9}$

$3\frac{2}{5} \times 1\frac{1}{3}$

$\frac{9}{11} \times 3\frac{1}{2}$

$12 \times 4\frac{8}{9}$

$1\frac{2}{5} \times 3\frac{1}{7}$

$2\frac{4}{9} \times 1\frac{7}{8}$

$1\frac{9}{10} \times 2\frac{1}{3}$

$4\frac{3}{4} \times 2\frac{2}{5}$

$1\frac{7}{8} \times 1\frac{4}{9}$

c

$1\frac{5}{6} \times 2\frac{3}{7}$

$4\frac{1}{4} \times 1\frac{1}{10}$

$1\frac{7}{8} \times 2\frac{1}{3}$

$3\frac{2}{3} \times 5\frac{1}{8}$

$2\frac{4}{7} \times 9$

$\frac{3}{8} \times 2\frac{1}{7}$

$4\frac{1}{5} \times 1\frac{3}{8}$

$7\frac{1}{3} \times \frac{3}{10}$

$2\frac{4}{5} \times 16$

How many ladders did you climb?

Divide fractions

Remember

When mentally dividing fractions, use the term 'how many' for the division symbol.

For example, $2\frac{3}{4} \div \frac{1}{4}$ is the same as $2\frac{3}{4}$, how many quarters? (11)

1 Divide these fractions in your head.

a	$3\frac{1}{2} \div \frac{1}{2}$	**b**	$4\frac{2}{3} \div \frac{1}{3}$	**c**	$1\frac{7}{9} \div \frac{1}{9}$	**d**	$8\frac{1}{4} \div \frac{1}{4}$	**e**	$2\frac{1}{8} \div \frac{1}{8}$	**f**	$5\frac{3}{4} \div \frac{1}{4}$
g	$3\frac{1}{5} \div \frac{1}{5}$	**h**	$6\frac{2}{7} \div \frac{1}{7}$	**i**	$1\frac{1}{2} \div \frac{1}{4}$	**j**	$2\frac{1}{4} \div \frac{1}{8}$	**k**	$6\frac{1}{2} \div \frac{1}{4}$	**l**	$3\frac{3}{4} \div \frac{1}{8}$
m	$1\frac{1}{3} \div \frac{1}{6}$	**n**	$3\frac{2}{3} \div \frac{1}{6}$	**o**	$2\frac{1}{5} \div \frac{1}{10}$	**p**	$4\frac{3}{5} \div \frac{1}{10}$	**q**	$4\frac{1}{4} \div \frac{1}{8}$	**r**	$3\frac{1}{2} \div \frac{1}{8}$
s	$5\frac{1}{2} \div \frac{1}{6}$	**t**	$2\frac{2}{3} \div \frac{1}{9}$	**u**	$4\frac{1}{3} \div \frac{1}{9}$	**v**	$2\frac{1}{2} \div \frac{1}{6}$	**w**	$1\frac{5}{6} \div \frac{1}{12}$	**x**	$3\frac{1}{6} \div \frac{1}{12}$

Remember

Where the divisor or dividend is a whole number, write the whole number as a fraction over 1.

2 Calculate each division and write the answer in its simplest form.

a	$2\frac{3}{5} \div 8$	**b**	$7\frac{2}{3} \div 1\frac{1}{8}$	**c**	$3\frac{4}{7} \div \frac{2}{3}$	**d**	$\frac{9}{10} \div \frac{3}{8}$	**e**	$4\frac{1}{8} \div 2\frac{3}{4}$	**f**	$1\frac{7}{8} \div 5$
g	$2\frac{1}{6} \div \frac{3}{10}$	**h**	$3\frac{4}{5} \div 1\frac{1}{3}$	**i**	$\frac{8}{9} \div 1\frac{1}{2}$	**j**	$3\frac{2}{5} \div 2\frac{1}{7}$	**k**	$5\frac{5}{9} \div 3\frac{1}{4}$	**l**	$12 \div 2\frac{2}{3}$
m	$3\frac{7}{10} \div 1\frac{5}{6}$	**n**	$7\frac{3}{4} \div \frac{4}{11}$	**o**	$4\frac{1}{9} \div 2\frac{1}{2}$	**p**	$5\frac{7}{8} \div \frac{4}{5}$	**q**	$15 \div 4\frac{1}{3}$	**r**	$\frac{8}{11} \div \frac{5}{9}$
s	$1\frac{6}{7} \div 2\frac{1}{4}$	**t**	$3\frac{11}{12} \div 1\frac{2}{3}$	**u**	$2\frac{5}{7} \div 1\frac{1}{9}$	**v**	$8\frac{7}{15} \div 9$	**w**	$4\frac{1}{12} \div \frac{3}{5}$	**x**	$2\frac{3}{10} \div 4\frac{1}{2}$

3 Copy and complete each circle by dividing the fractions around the outside by the fraction in the centre.

a

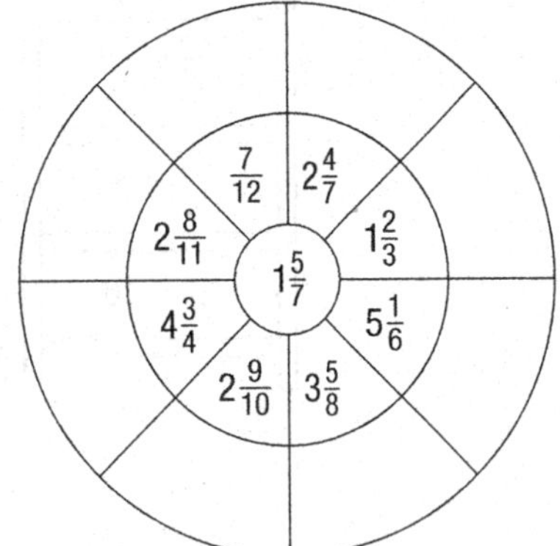

b

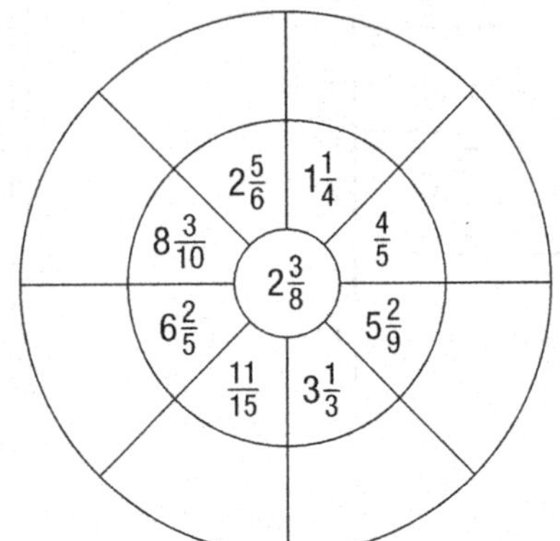

4 Divide these fractions. Write all answers in their simplest form.

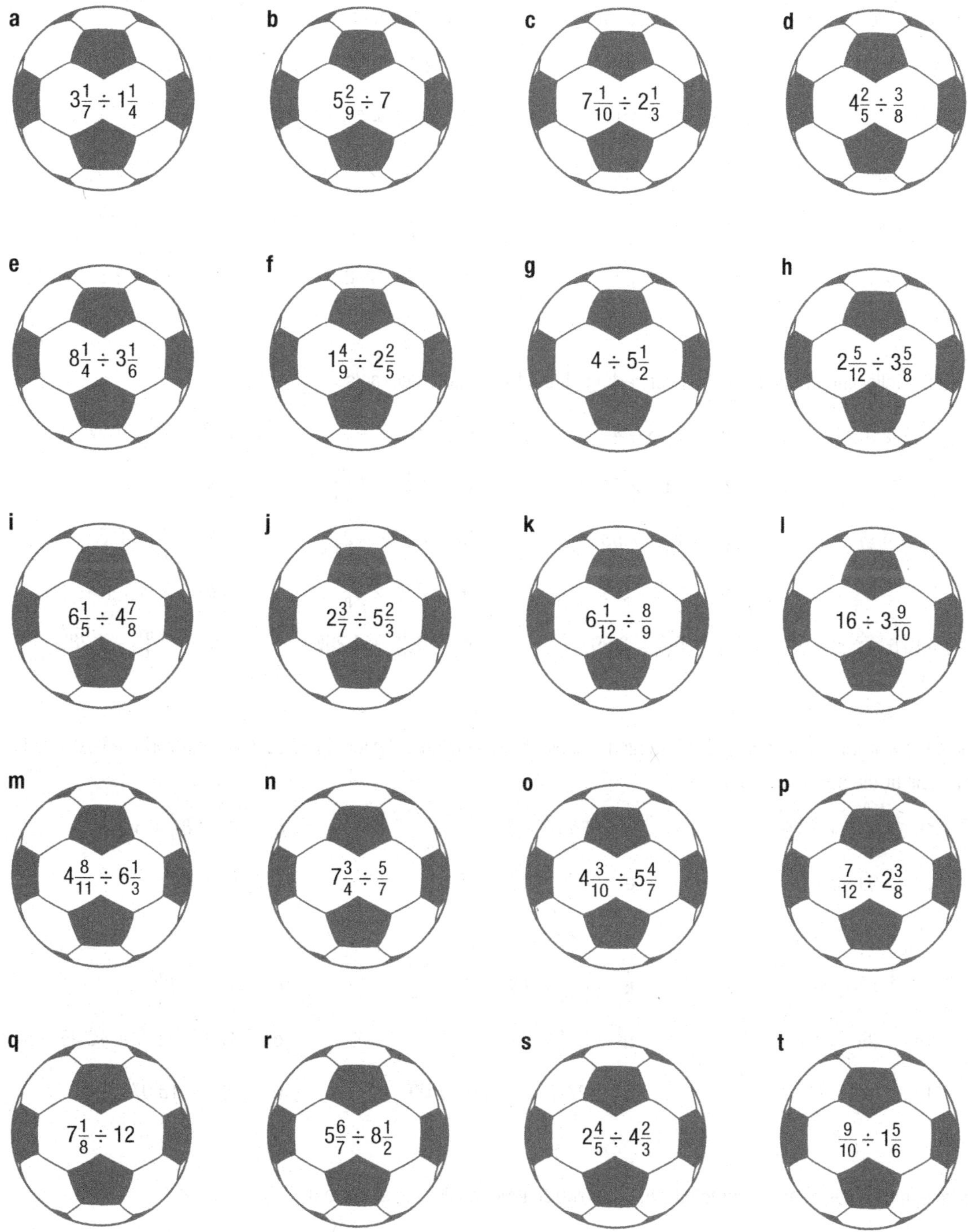

How many balls did you kick?

Use the four processes with fractions and decimals

1 Solve the following problems by first converting all the decimals to common fractions.

a	$8\frac{3}{4} + 1.75$	**b**	$7.5 - 4\frac{3}{8}$	**c**	$5\frac{2}{3} \div 0.5$	**d**	$2.7 \times 1\frac{1}{4}$
e	$5\frac{1}{2} \div 0.3$	**f**	$3\frac{1}{7} \times 4.1$	**g**	$3.25 + 2\frac{4}{9}$	**h**	$6\frac{2}{5} - 1.8$
i	$1.9 \times 5\frac{2}{11}$	**j**	$4\frac{5}{8} + 2.7$	**k**	$5.1 - 2\frac{1}{3}$	**l**	$7.75 \div 3\frac{5}{7}$
m	$7.3 - 4\frac{7}{8}$	**n**	$3.7 \div 1\frac{1}{6}$	**o**	$2\frac{5}{8} \times 3.4$	**p**	$5.6 + 4\frac{5}{6}$
q	$4\frac{3}{7} \div 2.15$	**r**	$2.24 \times 1\frac{3}{5}$	**s**	$7\frac{3}{4} + 3.08$	**t**	$6\frac{5}{7} - 2.8$

2 Solve the following problems by first converting all the fractions to decimals.

a	$3\frac{1}{2} + 2.9$	**b**	$6.75 - 4\frac{3}{10}$	**c**	$4.125 \div \frac{1}{4}$	**d**	$2\frac{3}{4} \times 1.57$
e	$\frac{5}{8} \div 5$	**f**	$4.08 \times 2\frac{4}{5}$	**g**	$8.14 + 5\frac{67}{100}$	**h**	$7\frac{9}{100} - 0.83$
i	$1\frac{2}{5} \times 4.87$	**j**	$1\frac{3}{8} + 7.53$	**k**	$6.172 - 2\frac{8}{100}$	**l**	$4\frac{1}{4} \div 0.5$
m	$9.23 - 4\frac{4}{5}$	**n**	$3\frac{6}{10} \div 0.3$	**o**	$2.67 \times 3\frac{1}{8}$	**p**	$2.081 + 4\frac{33}{100}$
q	$4.875 \div \frac{3}{4}$	**r**	$5\frac{9}{10} \times 2.89$	**s**	$6\frac{8}{100} + 4.267$	**t**	$7\frac{51}{100} - 1.285$

3 Solve the following additions and subtractions. Look at each one and decide if it is easier to convert the fractions to decimals or the decimals to fractions.

a	$15.6 + 7\frac{7}{10} + 6.29$	**b**	$54.07 - 23\frac{3}{4}$	**c**	$4\frac{5}{7} + 8\frac{2}{3} + 5.25$
d	$18\frac{2}{5} - 9.664$	**e**	$8\frac{3}{8} + 11\frac{3}{4} + 7.081$	**f**	$9\frac{5}{6} - 6.8$
g	$12\frac{2}{9} + 0.625 + 5\frac{1}{3}$	**h**	$22.835 - 13\frac{3}{5}$	**i**	$16.09 + 8.255 + 10\frac{57}{100}$
j	$14\frac{33}{100} - 7.816$	**k**	$6\frac{7}{12} + 3.7 + 8\frac{1}{4}$	**l**	$18.5 - 9\frac{5}{9}$
m	$19\frac{1}{6} + 25\frac{7}{8} + 13.4$	**n**	$31\frac{1}{4} - 23.807$	**o**	$15.85 + 8\frac{7}{15} + 12.35$
p	$17.8 - 8\frac{4}{9}$	**q**	$27\frac{7}{10} + 18.43 + 7.09$	**r**	$22\frac{2}{3} - 16.07$

4 Write some additions and subtractions of your own that are easier to calculate by:

a converting the fractions to decimals

b converting the decimals to fractions

Solve fraction word problems

Decide which process to use and then solve each word problem.

1 A builder starts a job with $1\frac{5}{8}$ kilograms of nails and when he finishes he has $\frac{2}{3}$ kilogram left. How many kilograms of nails did he use on the job?

2 Mary has a piece of fabric that is $5\frac{3}{8}$ metres long. She needs $1\frac{1}{7}$ metres to make a top.

- **a** How many tops can she make from her piece of fabric?
- **b** How much fabric will she have left over?

3 In a village of 126 people, 54 are children. Of these children, $\frac{1}{3}$ are too young for school and $\frac{2}{9}$ are in grade 6 or higher. The rest are in grades 1 to 5. Of the adults, $\frac{5}{8}$ are males.

- **a** What fraction of the people in the village are children?
- **b** How many children are too young for school?
- **c** How many children are in grades 1 to 5?
- **d** How many of the adults are females?

4 Samson needs $9\frac{4}{5}$ L of fertiliser for his garden. Lydia's garden is $2\frac{1}{3}$ the size of Samson's garden. How much fertiliser will she need?

5 Peter has $8\frac{3}{4}$ L of paint. He calculates that he needs $25\frac{2}{5}$ L of paint to paint a house. How much more paint does he need to buy?

6 A concert hall holds 720 people when full.

- **a** When the hall is $\frac{5}{8}$ full, how many people are there?
- **b** When the hall is $\frac{3}{4}$ full, how many seats are empty?
- **c** When the hall is $\frac{5}{6}$ full, how many people are there?
- **d** If the hall is full and $\frac{2}{9}$ of the people leave before the concert finishes, how many people are left in the hall?

7 An athlete trains for $2\frac{5}{12}$ hours each day. For how many hours does the athlete train in:

a 5 days? **b** 11 days? **c** 28 days? **d** 16 days?

8 A bucket holds $9\frac{4}{5}$ L of water. How many bottles, each holding $\frac{2}{3}$ L, can be filled from the bucket?

9 A large storage container has $15\frac{3}{8}$ kg of rice in it.

- **a** If another $8\frac{1}{5}$ kg is added to it, how much will it contain?
- **b** If $2\frac{3}{4}$ kg of rice is used, how much will still be in the container?

10 Five sacks are found to have a total weight of $114\frac{2}{3}$ kg. What is the average weight of each sack?

11 How many scarves, each $1\frac{2}{5}$ metres long, can be made from a roll of material that is $27\frac{2}{3}$ metres long?

12 In a swimming squad of 56 students, there are 32 girls. Of the students, $\frac{3}{8}$ of the girls and $\frac{5}{12}$ of the boys are training for the backstroke events.

- **a** How many students are training for backstroke?
- **b** If $\frac{1}{4}$ of the remaining girls are training for butterfly events, how many is this?
- **c** If $\frac{1}{2}$ of the remaining boys are training for breaststroke events, how many is this?

13 If a person walks at a rate of $6\frac{1}{3}$ km per hour, how many kilometres will she walk in:

- **a** $2\frac{1}{4}$ hours?
- **b** $1\frac{2}{5}$ hours?
- **c** $\frac{7}{10}$ hour?
- **d** $3\frac{5}{6}$ hours?

14 If $15\frac{9}{10}$ kg of tins are packed onto a table in the morning and another $12\frac{2}{3}$ kg of tins are packed onto the same table in the afternoon, what is the combined weight of tins on the table?

15 Isaac has $14\frac{2}{5}$ square metres of land for a garden. He plants beans in $\frac{1}{4}$ of his garden and cabbages in $\frac{5}{12}$ of it.

- **a** How many square metres is planted with beans?
- **b** How many square metres is planted with cabbages?
- **c** How many square metres does Isaac have left in which to plant other crops?

16 A carpenter calculated that the average length of each piece of wood in a stack was $1\frac{3}{20}$ metres. If there were 15 pieces of wood in the stack and they were placed end to end, how far would the row of pieces reach?

17 A shop that made hot chips cooked $14\frac{3}{5}$ kg of potatoes on Friday night, $19\frac{5}{8}$ kg on Saturday night and $16\frac{3}{4}$ kg on Sunday night.

- **a** How many kilograms of potatoes were cooked over the three nights?
- **b** How many more potatoes were cooked on Saturday night than on Friday night?
- **c** How many more potatoes were cooked on Saturday night than on Sunday night?
- **d** What was the average weight of potatoes cooked each of the three nights?

18 Sam sets out on his bike to cycle to the next village. He travels $27\frac{2}{5}$ km and still has another $18\frac{3}{4}$ km to travel.

- **a** How far apart are the two villages?
- **b** If Sam decides to return on the same day, how many kilometres will he travel altogether?

8.1.2 Use decimals to solve real life problems

Add and subtract decimals

1 When Grace went shopping she kept the dockets from each store.

SAM'S STORE	SUPER MART	SAVE MORE
K5.85	K11.45	K8.15
K4.50	K9.70	K17.30
K0.65	K3.55	K9.25
K1.35	K15.30	K11.60
K8.75	K5.20	K4.35
K2.95	K8.85	K1.05
K2.15	K7.35	K5.85
K3.80	K12.65	K3.70
K7.85	K7.50	K10.20
	K5.85	K8.90
	K2.45	
	K7.75	

a Find the total of each shopping docket.

b How much more money did Grace spend at Save More than at Sam's Store?

c How much more money did Grace spend at Super Mart than at Save More?

d What was the total amount of money spent by Grace at the three stores?

e What is the difference in price between the most expensive item and the least expensive item on the lists?

2 Florence visited two stores to find the price of furniture for her new house. The prices below show the cost of similar items at each store.

	STORE A	STORE B
Lounge suite	K1475.50	K1250.00
Dining table	K511.75	K549.95
Dining chairs (6)	K539.40	K513.00
Floor rug	K125.75	K109.80
Coffee table	K274.00	K298.95
Bookcase	K815.50	K799.95

a If Florence buys all her furniture from Store A, how much will she pay?

b Is this more or less than Store B? How much more or less?

c If Florence buys her lounge suite and floor rug from Store B and all her other furniture from Store A, how much will she pay in total?

d If she buys the less expensive items from either store, how much will she pay?

e If she buys the most expensive items from either store, how much will she pay?

f If Florence decides to just buy a lounge suite, bookcase and coffee table from Store B, how much change will she have from K2500?

3 Copy and complete each table.

a

+	7.08	34.64	16.7	281.5
153.7				
28.95				
9.068				
66.33				

b

−	58.03	4.672	132.7	23.14
25.6				
372.14				
85.69				
7.083				

c

+	29.344	76.8	153.2	51.6
88.516				
6.787				
58.12				
345.8				

d

−	82.3	56.17	5.79	164.4
17.09				
244.6				
75.83				
8.774				

e

+	84.04	6.885	66.7	186.9
7.067				
415.8				
98.54				
125.6				

f

−	243.8	46.7	74.18	155.72
62.09				
112.9				
36.4				
217.85				

4 Copy and complete these *difference* pyramids. For example:

			3.5			
		8.6		5.1		
	12.4		3.8		8.9	
20.1		7.7		11.5		2.6

a

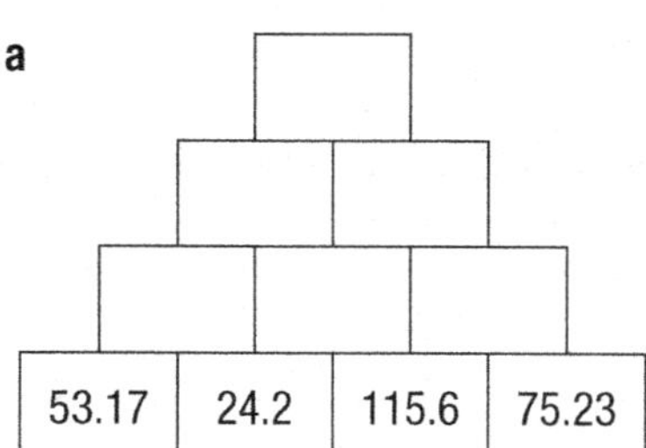

b

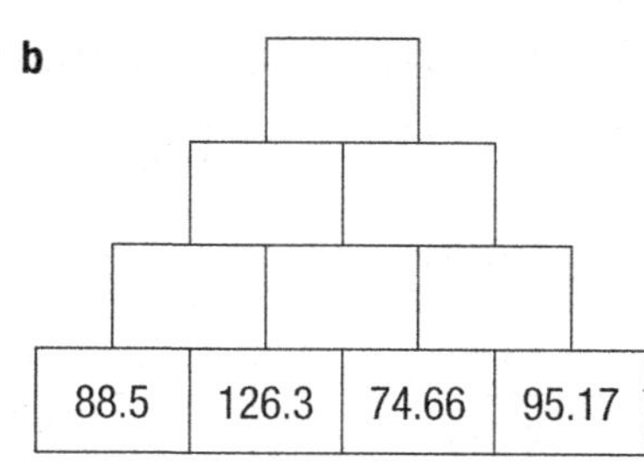

c

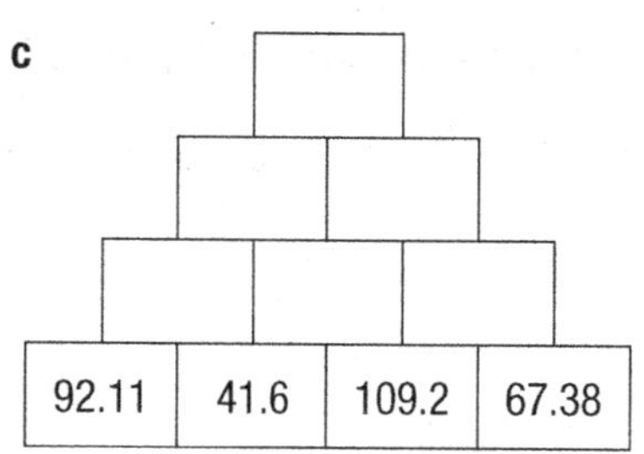

d

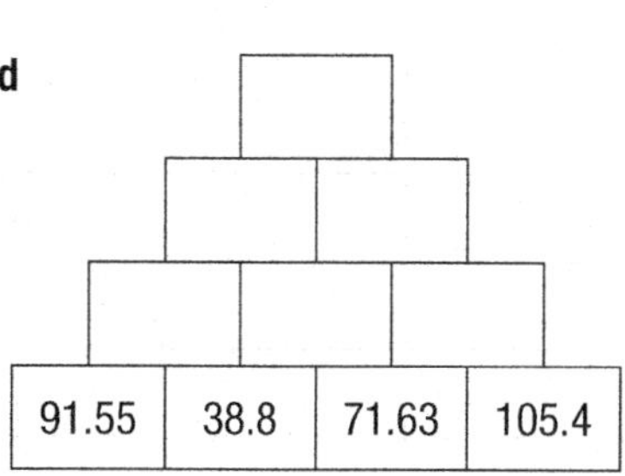

e

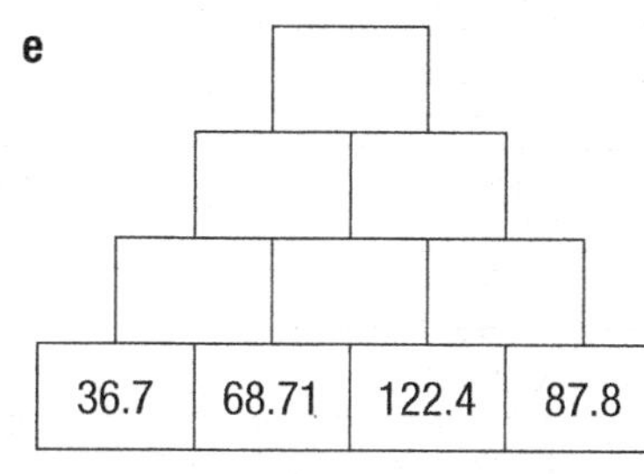

f

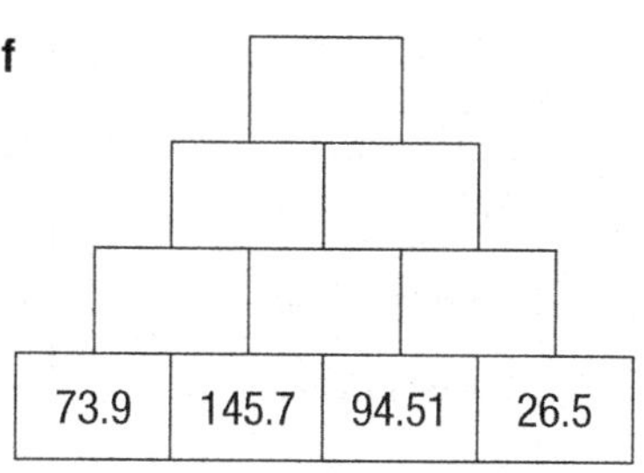

5 Copy and complete these *addition* pyramids. For example:

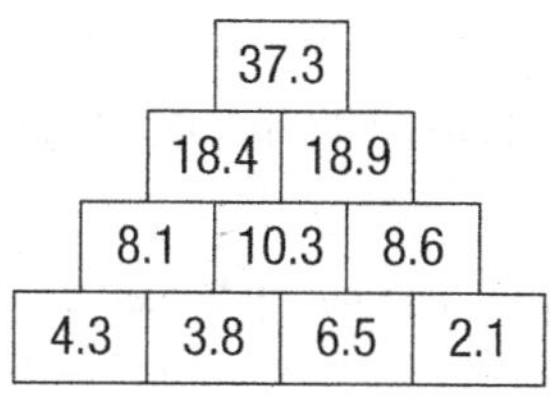

a

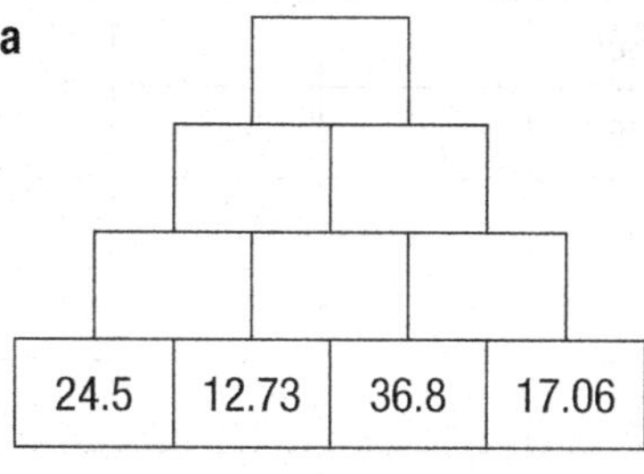

b

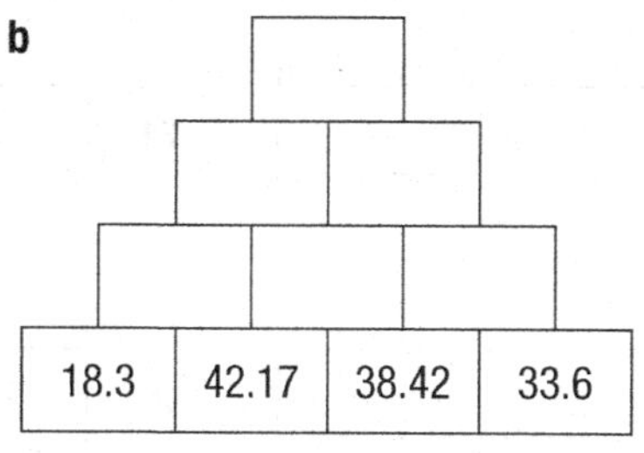

c

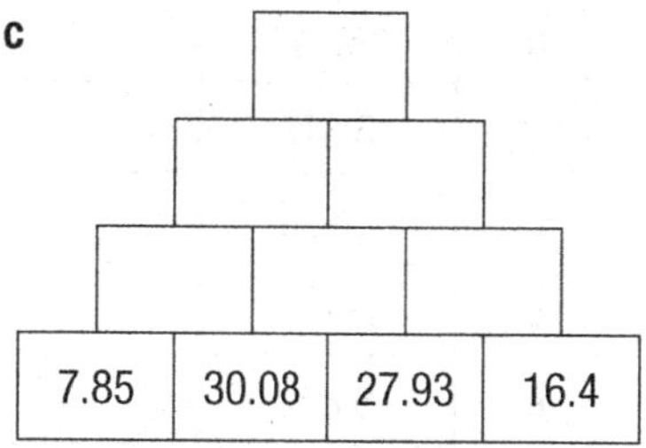

d

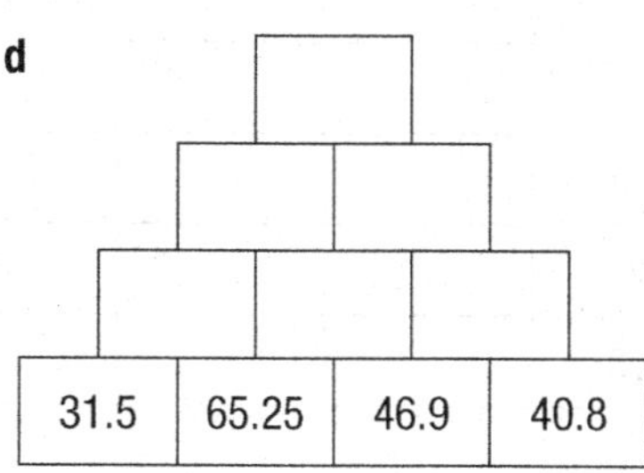

e

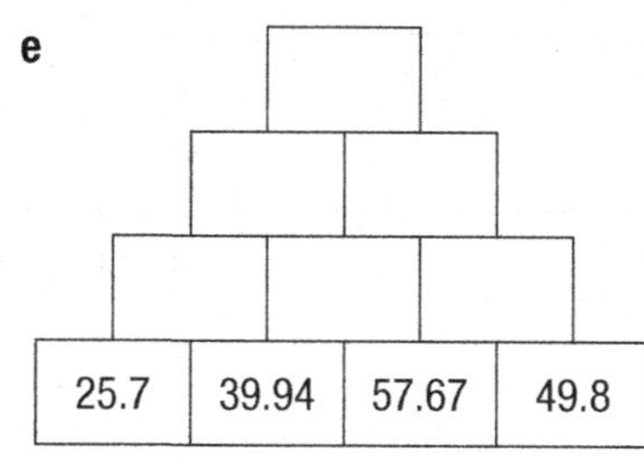

f

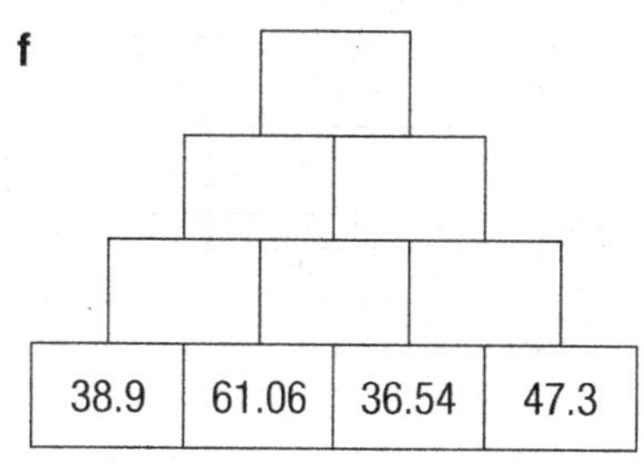

Multiply decimals by whole numbers and by 10, 100 and 1000

1 A tour leader wrote the following list of items and numbers of each item that he needed to buy for future tour groups.

Item	Cost per item	Number of item	Total cost
Coffee	K8.35	5	
Noodles	K1.20	25	
Rice	K12.45	8	
Tea	K6.80	4	
Sugar	K9.25	3	
Salt	K2.75	2	
Powdered milk	K5.50	6	
Sweet biscuits	K2.95	16	
Dry biscuits	K1.85	23	
Tomato sauce	K3.60	7	

a Use the 'Cost per item' and the 'Number of item' columns to calculate the total cost of each group of items.

b How much change would the tour leader get if he gave K500?

2 In each table, multiply the number on the left by 10, 100 and 1000.

	Number	× 10	× 100	× 1000
a	16.744			
b	5.817			
c	64.28			
d	431.77			
e	6.892			
f	3.15			
g	266.4			
h	0.573			
i	19.08			
j	31.097			
k	11.6			

	Number	× 10	× 100	× 1000
l	5.865			
m	24.282			
n	73.9			
o	44.76			
p	156.3			
q	9.066			
r	75.18			
s	3.7			
t	0.08			
u	343.9			
v	8.13			

Multiply decimals by multiples of 10

Remember

When multiplying decimals by multiples of 10, first multiply by 10 and then multiply by the number of tens.

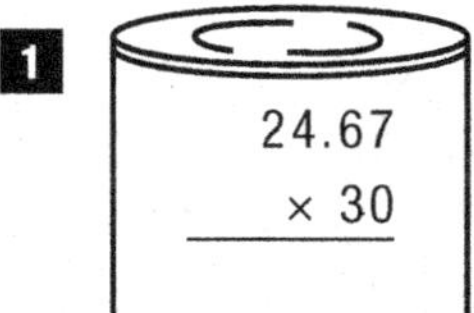

Multiply each decimal number by the given multiple of 10 to see how many tins you can push off the shelves.

1 24.67 × 30

2 163.08 × 60

3 9.553 × 20

4 47.6 × 80

5 235.9 × 50

6 68.74 × 70

7 8.494 × 40

8 108.75 × 90

9 74.66 × 20

10 8.976 × 80

11 328.7 × 50

12 93.4 × 30

13 244.8 × 90

14 54.718 × 60

15 93.45 × 40

16 168.7 × 70

17 3.614 × 80

18 179.8 × 50

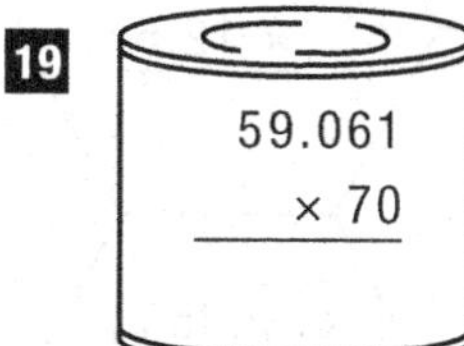

19 59.061 × 70

20 205.7 × 30

21 88.63 × 90

22 72.99 × 40

23 354.6 × 20

24 274.35 × 60

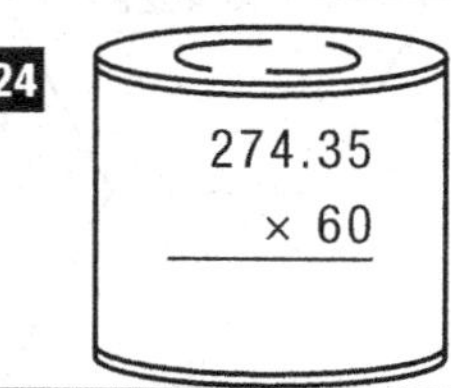

Multiply decimals by decimals

Remember

When multiplying decimals by decimals, first estimate the answer so you know where to place the decimal point.

Solve each multiplication to reach the honey in the centre of the honeycomb.

1

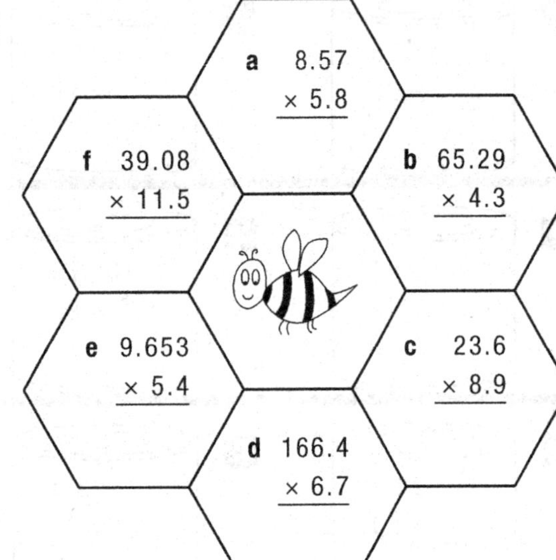

2

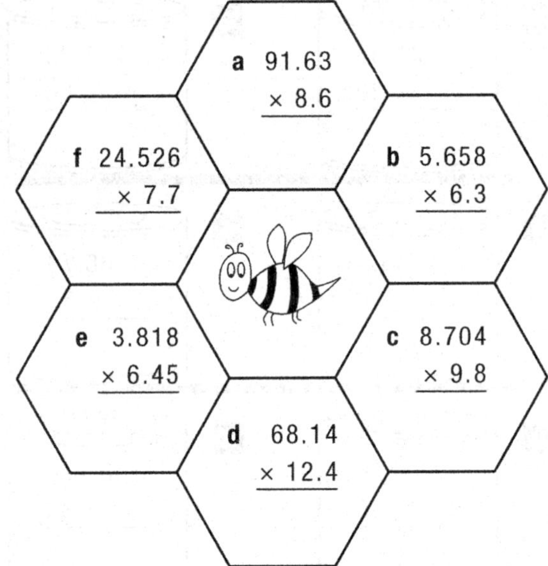

3

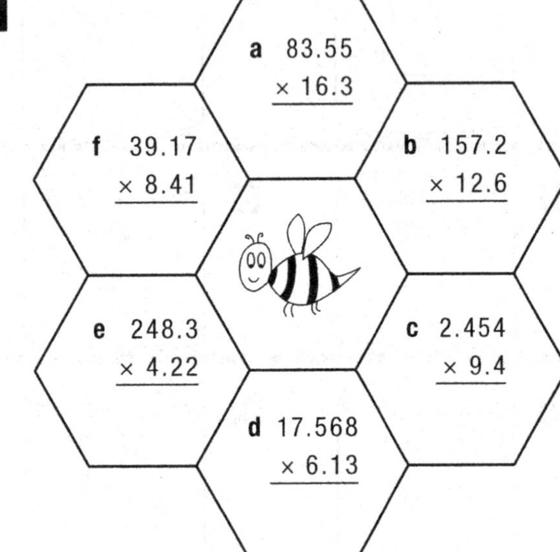

4

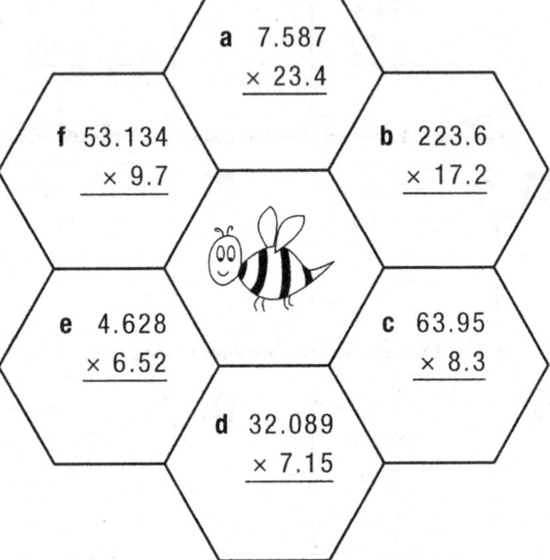

Divide decimals by 10, 100, 1000 and by multiples of 10

1 Divide the number on the left by 10, 100 and 1000.

	Number	÷ 10	÷ 100	÷ 1000
a	123.67			
b	174.3			
c	2775.9			
d	362.8			
e	215.4			
f	86.7			
g	491.08			
h	63.44			
i	118.5			
j	4216.3			
k	29.7			

	Number	÷ 10	÷ 100	÷ 1000
l	7.85			
m	53.6			
n	321.7			
o	2.48			
p	80.06			
q	9.004			
r	3.662			
s	0.59			
t	4.3			
u	92.18			
v	77.2			

Remember

When dividing decimals by multiples of 10, first divide by 10 and then divide by the number of tens.

2 Copy and complete each table.

a

÷ 30					
201.9	244.8	157.5	374.1	296.4	338.7

b

÷ 80					
389.6	164.8	683.2	1226.4	965.6	1378.4

c

÷ 40					
658.4	381.6	854.4	758.8	462.4	1115.6

d

÷ 70					
548.8	656.6	872.9	572.6	1370.6	1150.1

Divide decimals by whole numbers

Divide the decimals around each wheel by the number in the centre of the wheel.

1

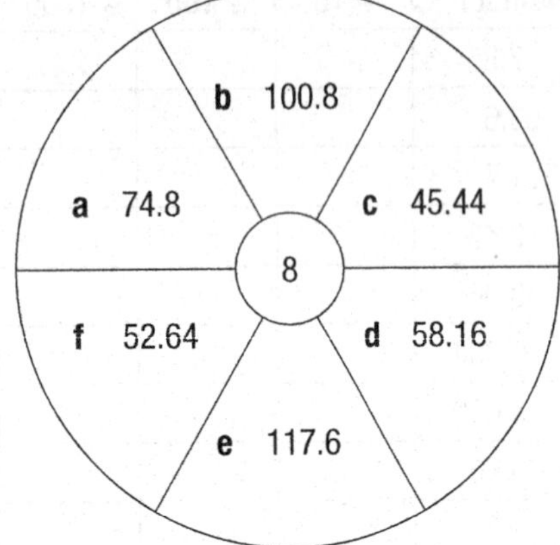

2

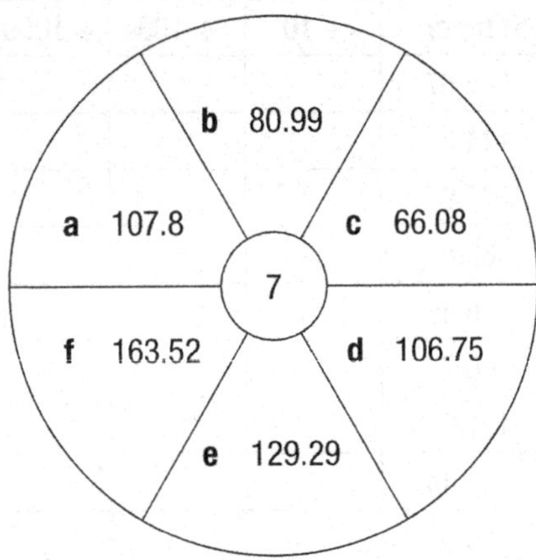

3

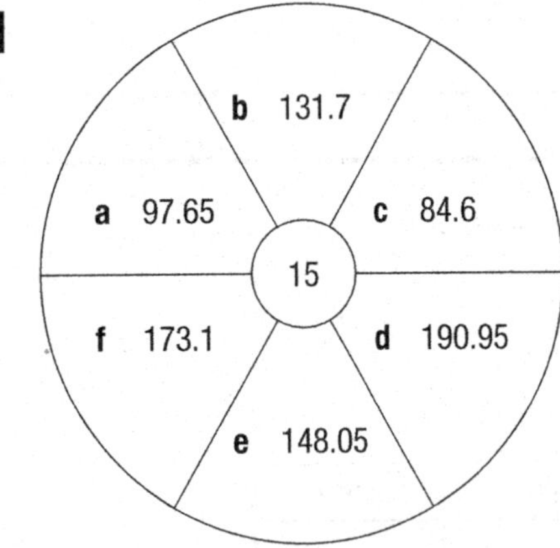

4

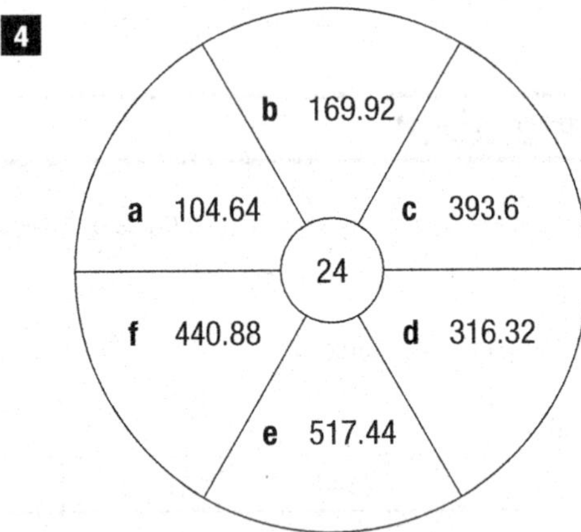

5

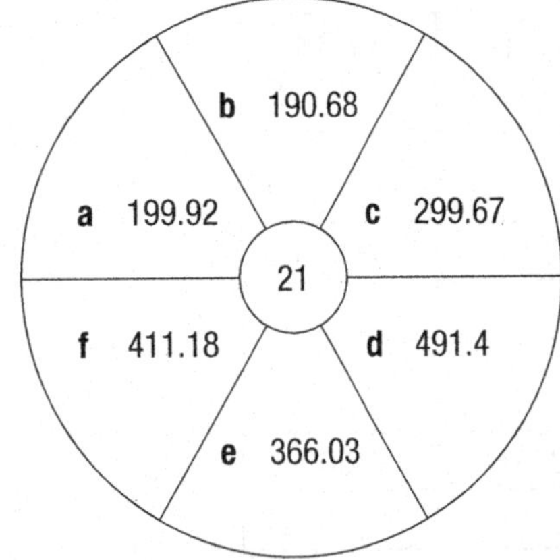

6

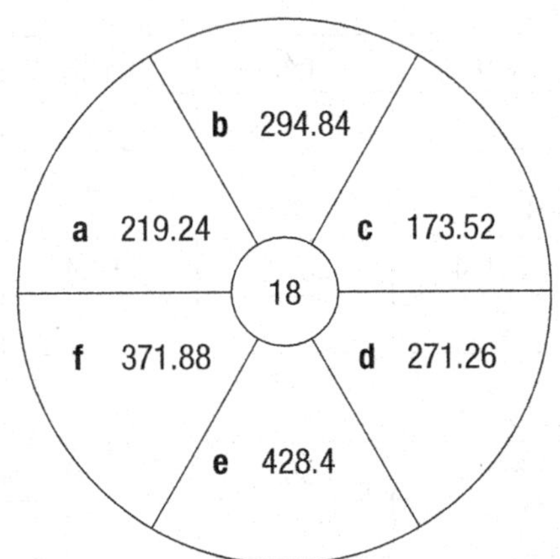

Divide decimals by decimals

Remember

When dividing decimals by decimals, first estimate the answer so you know where to place the decimal point.

How many coconuts can you crack? Calculate each answer to two decimal places.

1  $32.16 \div 4.8$

2 $27.26 \div 2.9$

3  $96.28 \div 8.3$

4 $104.52 \div 7.8$

5  $63.72 \div 3.6$

6 $87.18 \div 4.7$

7  $73.16 \div 8.2$

8 $167.58 \div 2.3$

9  $142.14 \div 5.4$

10 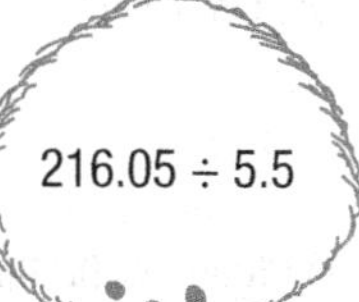 $216.05 \div 5.5$

11 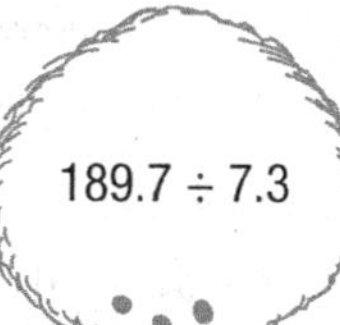 $189.7 \div 7.3$

12  $156.2 \div 4.4$

13 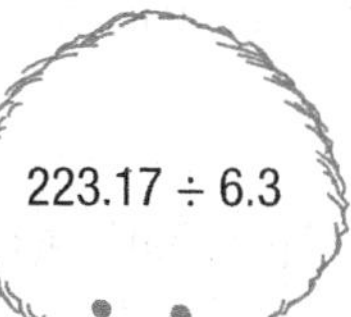 $223.17 \div 6.3$

14  $138.09 \div 3.9$

15  $96.213 \div 4.2$

16 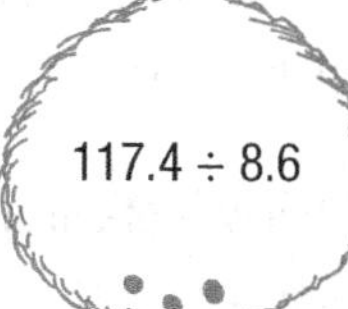 $117.4 \div 8.6$

17 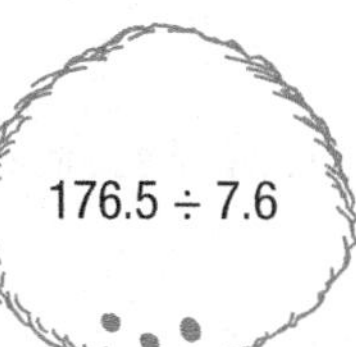 $176.5 \div 7.6$

18  $238.3 \div 9.1$

19  $221.14 \div 5.5$

20 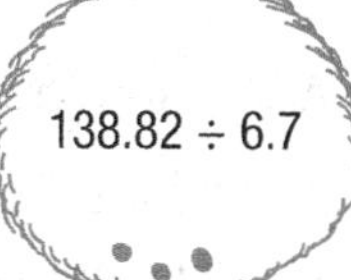 $138.82 \div 6.7$

Solve decimal word problems

Decide which process to use and then solve each word problem. Where appropriate, round answers to two decimal places.

1 How much would the following products cost?

- **a** 2.7 kg of mince beef at K8.65 per kg and 3.4 kg of potatoes at K2.95 per kg
- **b** 1.6 kg of bananas at K1.80 per kg and 2.3 kg of sausages at K5.75 per kg
- **c** 4.25 kg of chicken wings at K3.35 per kg and 2.8 kg of onions at K2.55 per kg

2 At a market, bilums sell for various prices. How much will I pay if I buy the following bilums?

- **a** 5 at K28.50 and 3 at K43.80
- **b** 2 at K31.60 and 6 at K35.75
- **c** 4 at K23.35 and 4 at K47.65
- **d** 3 at K29.45 and 5 at K38.95

3 A 30-metre length of rope is cut into three pieces. One piece is 11.38 m and another piece is 8.74 m. Find the length of the third piece.

4 How many 2.7 m lengths of fabric can be cut from these fabric lengths?

a 126.9 m **b** 89.1 m **c** 102.6 m **d** 51.3 m

5 A long car journey of 953.4 kilometres was spread over three days. On the first day, 274.8 kilometres were travelled. On the second day, 358.5 kilometres were travelled.

- **a** How many kilometres were travelled on the third day?
- **b** How many more kilometres were travelled on the second day than on the third day?
- **c** What was the average distance travelled each day?

6 A small water tank holds 537.5 L. After some water was used for the garden, there were 284.36 L left in the tank. How much water was used on the garden?

7 At a weight-lifting competition a weight lifter had four turns and managed to lift 144.7 kg, 152.3 kg, 147.9 kg and 156.4 kg.

- **a** What is the difference between his best and worst lifts?
- **b** What is the average of his four lifts?

8 At a gymnastics competition, a competitor scored 9.007, 8.95, 9.122 and 8.965 for her events and came second.

- **a** What is her combined score?
- **b** If the winner had a combined score of 36.51, what was the difference between first and second?
- **c** If the third placed competitor scored 35.876, what was the difference between second and third?

8.1.3 Convert freely between fractions, decimals, percentages and ratios

Convert between fractions and decimals

1 Write these decimals as common fractions in their simplest form.

a	0.25	**b**	0.8	**c**	0.48	**d**	0.35	**e**	0.16	**f**	0.62	**g**	0.54	**h**	0.12
i	0.04	**j**	0.77	**k**	0.85	**l**	0.28	**m**	0.188	**n**	0.255	**o**	0.072	**p**	0.375
q	0.625	**r**	0.06	**s**	0.32	**t**	0.4	**u**	0.96	**v**	0.44	**w**	0.66	**x**	0.248

2 Write these decimals as mixed numbers with the fraction in its simplest form.

a	2.6	**b**	7.05	**c**	12.38	**d**	5.125	**e**	9.062	**f**	15.75
g	8.148	**h**	4.56	**i**	11.006	**j**	25.18	**k**	14.625	**l**	6.464
m	10.02	**n**	8.076	**o**	5.875	**p**	3.26	**q**	7.75	**r**	9.866
s	4.078	**t**	7.155	**u**	10.42	**v**	8.34	**w**	11.88	**x**	2.512

3 Write these fractions as decimals.

a	$\frac{3}{10}$	**b**	$\frac{77}{100}$	**c**	$\frac{4}{5}$	**d**	$\frac{61}{100}$	**e**	$\frac{3}{20}$	**f**	$\frac{4}{100}$	**g**	$\frac{138}{1000}$	**h**	$\frac{513}{1000}$
i	$\frac{89}{1000}$	**j**	$\frac{16}{50}$	**k**	$\frac{217}{500}$	**l**	$\frac{18}{25}$	**m**	$\frac{63}{250}$	**n**	$\frac{78}{125}$	**o**	$\frac{11}{20}$	**p**	$\frac{119}{200}$
q	$\frac{28}{40}$	**r**	$\frac{7}{8}$	**s**	$\frac{14}{25}$	**t**	$\frac{36}{60}$	**u**	$\frac{133}{250}$	**v**	$\frac{12}{30}$	**w**	$\frac{53}{1000}$	**x**	$\frac{31}{50}$

4 Write these mixed numbers as decimals.

a	$2\frac{3}{5}$	**b**	$1\frac{9}{10}$	**c**	$16\frac{19}{100}$	**d**	$11\frac{6}{20}$	**e**	$6\frac{3}{4}$
f	$5\frac{1}{8}$	**g**	$10\frac{29}{50}$	**h**	$9\frac{13}{25}$	**i**	$25\frac{130}{1000}$	**j**	$8\frac{2}{5}$
k	$5\frac{63}{100}$	**l**	$12\frac{371}{1000}$	**m**	$7\frac{9}{30}$	**n**	$11\frac{29}{200}$	**o**	$4\frac{56}{250}$
p	$9\frac{11}{50}$	**q**	$3\frac{19}{25}$	**r**	$1\frac{9}{1000}$	**s**	$12\frac{42}{60}$	**t**	$8\frac{149}{500}$

Help Box

To convert a common fraction to a decimal, divide the numerator by the denominator.

Example: $\frac{5}{8} = 8\overline{)5.0}$

```
   0.625
8 )5.0
   4.8
    20
    16
     40
     40
      0
```

$\frac{5}{8} = 0.625$

A decimal that repeats is called a recurring or repeating decimal. For example, $\frac{1}{6}$ makes the recurring decimal 0.1666. We write this with a dot above the recurring numbers so 0.1666 becomes $0.1\dot{6}$.

5 Write each of these fractions as decimals (to three decimal places unless recurring).

a	$\frac{3}{8}$	**b**	$\frac{5}{12}$	**c**	$\frac{2}{7}$	**d**	$\frac{7}{9}$	**e**	$\frac{1}{6}$	**f**	$\frac{2}{3}$	**g**	$\frac{3}{11}$	**h**	$\frac{6}{7}$	**i**	$\frac{1}{12}$
j	$\frac{5}{6}$	**k**	$\frac{1}{9}$	**l**	$\frac{7}{11}$	**m**	$\frac{4}{9}$	**n**	$\frac{1}{3}$	**o**	$\frac{8}{11}$	**p**	$\frac{7}{12}$	**q**	$\frac{4}{7}$	**r**	$\frac{11}{15}$

Convert between fractions, decimals and percentages

Remember

To change a common fraction or a decimal to a percentage, multiply by 100.
Example: $\frac{5}{8} \times 100 = \frac{500}{8} = 62.5\%$ or $62\frac{1}{2}\%$; $0.27 \times 100 = 27\%$

To change a percentage to a common fraction or decimal, divide by 100.
Example: $38\% = \frac{38}{100} = \frac{19}{50}$ or $38 \div 100 = 0.38$

Copy and complete each chart. Write each fraction in its simplest form.

1

	Fraction	Decimal	Percentage
a	$\frac{3}{5}$		
b		0.71	
c			49%
d		0.42	
e	$\frac{1}{8}$		
f			28%
g	$\frac{11}{20}$		
h		0.03	
i		0.432	
j			53%
k	$\frac{17}{100}$		
l		0.55	
m			8%
n	$\frac{3}{8}$		
o			15%
p		0.675	
q	$\frac{7}{10}$		
r		0.824	
s			$17\frac{1}{2}\%$
t	$\frac{21}{50}$		
u		0.16	
v			96%
w		0.005	
x	$\frac{9}{25}$		

2

	Fraction	Decimal	Percentage
a		1.6	
b			227%
c	$1\frac{9}{10}$		
d	$3\frac{4}{10}$		
e		2.75	
f			114%
g		1.003	
h		5.4	
i	$2\frac{22}{25}$		
j			$12\frac{3}{4}\%$
k			$38\frac{1}{2}\%$
l		4.07	
m	$1\frac{17}{20}$		
n			156%
o	$2\frac{17}{50}$		
p		3.514	
q			$7\frac{1}{4}\%$
r		1.48	
s	$3\frac{4}{5}$		
t		2.183	
u	$5\frac{1}{2}$		
v			$23\frac{1}{2}\%$
w		5.22	
x	$7\frac{1}{5}$		

Convert between fractions, decimals, percentages and ratios

Remember

If a ratio contains only two terms, then it can be written as a fraction. For example, the ratio 2:7 has two terms (2 and 7) so can be written as $\frac{2}{7}$.

1 Change these ratios to fractions reduced to their simplest form. Where necessary, write fractions as mixed numbers.

a	6:10	**b**	3:4	**c**	8:12	**d**	6:4	**e**	10:6	**f**	3:9
g	12:4	**h**	5:6	**i**	7:14	**j**	8:3	**k**	11:5	**l**	2:7
m	6:15	**n**	4:3	**o**	9:8	**p**	3:5	**q**	8:10	**r**	15:24
s	7:3	**t**	13:9	**u**	4:30	**v**	10:7	**w**	12:18	**x**	9:15

2 Reduce these fractions to their simplest form and then write them as ratios. Change mixed numbers to improper fractions before writing as ratios.

a	$\frac{4}{10}$	**b**	$1\frac{3}{4}$	**c**	$\frac{2}{9}$	**d**	$\frac{10}{15}$	**e**	$2\frac{2}{5}$	**f**	$\frac{8}{12}$	**g**	$1\frac{3}{8}$	**h**	$2\frac{4}{7}$
i	$\frac{9}{21}$	**j**	$1\frac{7}{8}$	**k**	$\frac{15}{35}$	**l**	$\frac{3}{11}$	**m**	$3\frac{1}{4}$	**n**	$8\frac{1}{2}$	**o**	$\frac{4}{20}$	**p**	$2\frac{5}{7}$
q	$\frac{9}{30}$	**r**	$\frac{10}{7}$	**s**	$\frac{2}{16}$	**t**	$5\frac{1}{3}$	**u**	$1\frac{2}{3}$	**v**	$\frac{4}{9}$	**w**	$\frac{12}{40}$	**x**	$\frac{16}{34}$

3 Copy and complete the tables. Reduce all fractions to their simplest form.

a

Percentage	Decimal	Fraction	Ratio
35%			
		$\frac{9}{10}$	
	0.16		
			4:5
	0.07		

b

Percentage	Decimal	Fraction	Ratio
		$\frac{3}{25}$	
			12:16
48%			
	0.7		
			3:8

4 Write these in order from smallest to largest.

a	0.7, 1:4, 33%, $\frac{1}{2}$	**b**	7:5, $\frac{4}{5}$, 55%, 0.3	**c**	83%, 0.76, $\frac{3}{4}$, 24:25	**d**	6:11, 14%, $\frac{1}{3}$, 0.05
e	71%, 0.7, $\frac{21}{50}$, 1:2	**f**	$\frac{9}{10}$, 8:5, 85%, 1.2	**g**	$\frac{1}{8}$, 50%, 3:8, 0.8	**h**	3:2, $\frac{2}{3}$, 23%, 2.5
i	0.75, 36%, 2:4, $\frac{3}{10}$	**j**	99%, 0.9, $\frac{1}{9}$, 9:4	**k**	$\frac{1}{10}$, 4%, 0.4, 1:20	**l**	$\frac{1}{4}$, 4:1, 14%, 0.44
m	0.35, $\frac{3}{5}$, 4:5, 18%	**n**	57%, 5:3, $1\frac{1}{2}$, 1.25	**o**	150%, 1.3, $1\frac{1}{4}$, 3:1	**p**	$\frac{4}{25}$, 5%, 0.17, 2:25
q	0.83, $\frac{9}{10}$, 3:4, 81%	**r**	7:2, $2\frac{1}{2}$, 275%, 3.05	**s**	29%, 0.4, 3:10, $\frac{7}{20}$	**t**	5:7, 60%, 0.63, $\frac{9}{10}$
u	0.34, $\frac{21}{50}$, 1:5, 37%	**v**	1:3, $\frac{4}{9}$, 0.25, 18%	**w**	1.75, $1\frac{3}{5}$, 4:7, 100%	**x**	$\frac{7}{8}$, 8:7, 78%, 0.87

8.1.4 Solve problems in any situation that involves percentages

Express one number as a percentage of another

Remember

Make sure that both numbers are in the same unit of measurement before writing them as a fraction, then multiply by 100 and add the percentage sign.

For example, to express 45 toea as a percentage of K2, write both numbers as toea (45t out of 200t) and then write them as a fraction ($\frac{45}{200}$). Multiply $\frac{45}{200}$ by 100: $\frac{45}{200} \times \frac{100}{1} = 22.5\%$

1 Write these numbers as percentages. Round your answers to one decimal place.

a	27 out of 125	**b**	16 out of 86	**c**	48 out of 234	**d**	35 out of 154
e	12 out of 115	**f**	56 out of 345	**g**	44 out of 116	**h**	61 out of 500
i	36 out of 242	**j**	29 out of 184	**k**	45 out of 312	**l**	74 out of 526
m	52 out of 266	**n**	67 out of 453	**o**	23 out of 476	**p**	17 out of 322

2 Express the first amount as a percentage of the second. Change both amounts to the same unit where necessary, and round your answers to one decimal place.

a	600 g out of 1 kg	**b**	24t out of K3	**c**	K96 out of K125
d	73 cm out of 3 m	**e**	167 kg out of 550 kg	**f**	67 mm out of 11.5 cm
g	75 g out of 1.4 kg	**h**	115 mL out of 0.5 L	**i**	55t out of K2.45
j	430 m out of 1.6 km	**k**	545 kg out of 1.7 t	**l**	276 km out of 1240 km
m	77t out of K4.15	**n**	675 mL out of 1.85 L	**o**	235 g out of 2.3 kg

3 Write these larger numbers as percentages. Round your answers to one decimal place.

a	2420 out of 7360	**b**	1935 out of 4270	**c**	2780 out of 9125
d	1265 out of 6775	**e**	3450 out of 5835	**f**	1525 out of 7750
g	4285 out of 9000	**h**	2545 out of 3890	**i**	5110 out of 8480
j	6263 out of 9812	**k**	3472 out of 8164	**l**	2978 out of 7349
m	12 472 out of 20 756	**n**	9728 out of 19 886	**o**	6334 out of 21 058

4 A wealthy man died and left an amount of K125 000 to his relatives. The list below shows what each person received.

Wife K76 875	Daughter K20 500
Grandsons (3) K4500 each	Brother K14 125

What percentage of the total amount did each person receive?

Increase or decrease amounts by a certain percentage

Help Box

Multiplication can be used to increase or decrease a number by a certain percentage.

Example 1: To **increase** 85 by 10%, multiply 85 by 1.1 (100% + 10%). So, 85 × 1.1 = 93.5 or 85 increased by 10% is 93.5.

Example 2: To **decrease** 120 by 15%, multiply 120 by 0.85 (100% – 15%). So, 120 × 0.85 = 102 or 120 decreased by 15% is 102.

1 What would you multiply by to **increase** a number by the following percentages?

a	15%	**b**	20%	**c**	10%	**d**	30%	**e**	12%	**f**	25%
g	50%	**h**	5%	**i**	8%	**j**	$9\frac{1}{2}$%	**k**	22.5%	**l**	6.5%
m	11.2%	**n**	$15\frac{1}{2}$%	**o**	2.7%	**p**	26.8%	**q**	12.4%	**r**	$20\frac{1}{2}$%

2 What would you multiply by to **decrease** a number by the following percentages?

a	10%	**b**	50%	**c**	5%	**d**	35%	**e**	18%	**f**	15%
g	20%	**h**	30%	**i**	40%	**j**	75%	**k**	7%	**l**	25%
m	12.5%	**n**	5.5 %	**o**	$3\frac{1}{2}$%	**p**	$17\frac{1}{2}$%	**q**	8.2%	**r**	16.7%

3 Find these percentage increases.

a	50 by 25%	**b**	24 by 10%	**c**	12 by 30%	**d**	36 by 5%
e	27 by 20%	**f**	63 by 15%	**g**	19 by 7.5%	**h**	45 by $12\frac{1}{2}$%
i	9 by 10%	**j**	55 by 8.7%	**k**	21 by 5%	**l**	68 by $22\frac{1}{2}$%

4 Find these percentage decreases.

a	40 by 15%	**b**	38 by 10%	**c**	56 by 20%	**d**	26 by 30%
e	18 by 5%	**f**	20 by 1%	**g**	35 by 17%	**h**	64 by 25%
i	32 by 12%	**j**	50 by 6%	**k**	47 by 23%	**l**	75 by 15%

5 Increase these numbers by 15%.

a	105	**b**	250	**c**	128	**d**	312	**e**	236	**f**	460
g	345	**h**	184	**i**	210	**j**	475	**k**	507	**l**	386
m	277	**n**	408	**o**	586	**p**	162	**q**	329	**r**	298

6 Decrease these numbers by 7%.

a	175	**b**	320	**c**	246	**d**	378	**e**	182	**f**	516
g	432	**h**	109	**i**	355	**j**	290	**k**	569	**l**	141
m	254	**n**	518	**o**	163	**p**	488	**q**	399	**r**	236

Calculate percentage increase or decrease

Help Box

To calculate a percentage increase:

$$\frac{\text{increase}}{\text{original number}} \times \frac{100}{1}$$

So, an increase from 28 to 36

$= \frac{8}{28} \times \frac{100}{1} = \frac{800}{28} = 28.57\%$

Round 28.57% to one decimal place: 28.6%

There is a 28.6% increase from 28 to 36.

To calculate a percentage decrease:

$$\frac{\text{decrease}}{\text{original number}} \times \frac{100}{1}$$

So, a decrease from 55 to 42

$= \frac{13}{55} \times \frac{100}{1} = \frac{1300}{55} = 23.63\%$

Round 23.63% to one decimal place: 23.6%

There is a 23.6% decrease from 50 to 42.

1 Calculate the percentage increase when:

a	40 becomes 45	**b**	32 becomes 39	**c**	26 becomes 32
d	35 becomes 50	**e**	48 becomes 60	**f**	17 becomes 26
g	23 becomes 38	**h**	70 becomes 100	**i**	56 becomes 68
j	80 becomes 110	**k**	62 becomes 88	**l**	74 becomes 85
m	58 becomes 75	**n**	90 becomes 116	**o**	85 becomes 106

2 Calculate the percentage decrease when:

a	27 becomes 20	**b**	24 becomes 15	**c**	35 becomes 28
d	41 becomes 30	**e**	36 becomes 26	**f**	52 becomes 40
g	64 becomes 56	**h**	48 becomes 30	**i**	60 becomes 55
j	95 becomes 85	**k**	72 becomes 55	**l**	83 becomes 75
m	106 becomes 80	**n**	112 becomes 98	**o**	120 becomes 96

3 Calculate the percentage profit when a company increases its earnings from:

a	K10 500 to K15 000	**b**	K16 000 to K17 000	**c**	K12 300 to K14 800
d	K20 000 to K22 500	**e**	K18 200 to K23 500	**f**	K15 700 to K16 200
g	K21 900 to K24 350	**h**	K19 950 to K22 050	**i**	K17 450 to K20 100
j	K14 350 to K17 800	**k**	K24 700 to K26 250	**l**	K33 000 to K37 400

4 Calculate the percentage loss when a company decreases its earnings from:

a	K14 000 to K9500	**b**	K17 000 to K15 000	**c**	K12 500 to K11 800
d	K17 500 to K16 700	**e**	K18 200 to K17 000	**f**	K21 300 to K19 600
g	K24 650 to K23 000	**h**	K25 750 to K23 250	**i**	K32 000 to K27 450
j	K28 370 to K24 890	**k**	K31 090 to K28 420	**l**	K35 640 to K32 075

Calculate Value Added Tax (VAT)

Help Box

VAT is a 10% tax imposed on the sale of goods and services to raise revenue for the government. It is usually added on before a purchase price is given. When including VAT in the price of goods or services, it can be done in the same way as a 10% increase. For example, a service with a before-tax cost of K159 to which 10% VAT has to be added is the same as K159 × 1.1 = K174.90. So the purchase price is K174.90.

There are two methods to calculate the actual amount of VAT that is to be included.

1. Change 10% to a fraction and then multiply: $\frac{1}{10}$ × K159 = K15.90
2. Change 10% to its decimal form and then multiply: 10% of K159 = 0.1 × 159 = K15.90

1 The prices listed below do not include VAT. Find the price after VAT is included by multiplying each price by 1.1

a	K235	**b**	K127	**c**	K358	**d**	K146	**e**	K274	**f**	K453
g	K124.50	**h**	K93.50	**i**	K516	**j**	K281	**k**	K75.50	**l**	K309
m	K428	**n**	K366	**o**	K182.50	**p**	K220.50	**q**	K173.50	**r**	K245

2 Use one of the methods in the Help Box to calculate the amount of VAT (10%) that has to be added to each of these prices.

a	K416	**b**	K211	**c**	K749	**d**	K2018	**e**	K1382	**f**	K2887
g	K5706	**h**	K3649	**i**	K2817	**j**	K7294	**k**	K6552	**l**	K3918
m	K506.50	**n**	K373.50	**o**	K988.50	**p**	K239.50	**q**	K117.50	**r**	K465.50

Help Box

The amount of VAT that is included in a purchase price can be calculated by dividing by 11. For example, if a purchase price is K51.70 then K51.70 ÷ 11 = K4.70. So, the VAT is K4.70 and the price before VAT is K47 (K51.70 – K4.70).

3 Divide by 11 to calculate the amount of VAT that is included in each of these purchase prices.

a	K63.80	**b**	K83.60	**c**	K45.10	**d**	K41.80	**e**	K35.20	**f**	K57.20
g	K105.60	**h**	K93.50	**i**	K180.40	**j**	K169.40	**k**	K149.60	**l**	K133.10
m	K217.80	**n**	K179.30	**o**	K224.40	**p**	K343.20	**q**	K216.70	**r**	K258.50
s	K400.40	**t**	K353.10	**u**	K339.90	**v**	K456.50	**w**	K190.30	**x**	K284.90

4 Calculate the cost before tax on the following prices.

a	K47.30	**b**	K30.80	**c**	K72.60	**d**	K62.70	**e**	K42.90	**f**	K90.20
g	K106.70	**h**	K168.30	**i**	K146.30	**j**	K188.10	**k**	K137.50	**l**	K173.80
m	K225.50	**n**	K207.90	**o**	K237.60	**p**	K279.40	**q**	K314.60	**r**	K272.80
s	K178.20	**t**	K342.10	**u**	K459.80	**v**	K519.20	**w**	K487.30	**x**	K337.70

Calculate discounts, sale prices and commissions

Remember

A discount is a reduction in price, most often seen during a sale as a percentage, for example '20% off'.

A commission is a percentage paid to a person and is usually dependent on the amount of goods sold or business obtained. For example: one salesperson received a 7% commission on every computer he sold.

Copy and complete each chart by filling in the spaces.

1

	Price before discount	Discount	Price after discount
a	K45	15%	
b	K110	10%	
c	K84	15%	
d	K17	5%	
e	K145	20%	
f	K434	10%	
g	K238	$12\frac{1}{2}$%	
h	K87.50	20%	
i	K567.50	10%	
j	K188	25%	
k	K64	15%	
l	K209	30%	
m	K312.50	20%	
n	K132	$7\frac{1}{2}$%	
o	K383.60	25%	
p	K49.50	10%	
q	K258	$12\frac{1}{2}$%	
r	K775	15%	
s	K456.50	20%	
t	K96.50	30%	
u	K276	$7\frac{1}{2}$%	
v	K378	15%	

2

	Amount before commission	Commission	Amount after commission
a	K2300	5%	
b	K1450	7%	
c	K1875	10%	
d	K5250	8%	
e	K2765	5%	
f	K3425	3%	
g	K1910	12%	
h	K2475	10%	
i	K4800	$7\frac{1}{2}$%	
j	K3550	4%	
k	K2945	9%	
l	K1090	$6\frac{1}{2}$%	
m	K6745	8%	
n	K4125	12%	
o	K3220	5%	
p	K7500	$7\frac{1}{2}$%	
q	K2660	6%	
r	K3895	10%	
s	K8500	$5\frac{1}{2}$%	
t	K5870	11%	
u	K6430	7%	
v	K4990	$3\frac{1}{2}$%	

Calculate the full amount from a given percentage

Help Box

If a percentage of a full amount is known, then the full amount can be calculated by following the steps below. For example, if 80% of the full amount is K50, what is the full amount?

Step 1: Find 1% of the amount by dividing K50 by 80: K50 ÷ 80 = K0.625

Step 2: Find 100% of the amount by multiplying K0.625 by 100: K0.625 × 100 = K62.50

The full amount is K62.50.

1 The following amounts of money represent 80% of the full amount. Calculate 100% of each amount.

a	K60	b	K130	c	K45	d	K150	e	K90	f	K25
g	K175	h	K260	i	K110	j	K225	k	K320	l	K230
m	K124	n	K308	o	K256	p	K178	q	K296	r	K144

2 The following amounts of money represent 65% of the full amount. Calculate 100% of each amount.

a	K780	b	K585	c	K1365	d	K1170	e	K1690	f	K2730
g	K2405	h	K2275	i	K3445	j	K3055	k	K2535	l	K3640
m	K4745	n	K5915	o	K5200	p	K4420	q	K5785	r	K6370

3 Copy and complete this table. The first one is done for you.

	Full price	35% of full price	15% of full price	60% of full price	75% of full price	45% of full price
a	K2200	K770	K330	K1320	K1650	K990
b		K420				K540
c				K1020	K1275	
d			K510			
e		K875				
f				K2460		
g					K2775	
h						K1597.50
i			K637.50			
j		K2142				
k					K4230	
l				K1698		
m			K220.50			
n						K1476

Solve percentage word problems

1 If Daniel pays K1564 for an airfare and this is 85% of the full fare, how much is the full fare?

2 A water tank holds 1400 litres when full. When it is 65% full:

- **a** how much water is in the tank?
- **b** how much more water is needed until the tank is full?

3 A computer costs K1259 before VAT of 10% is added.

- **a** How much VAT has to be added?
- **b** What is the cost of the computer after VAT is added?

4 A holiday costs K2695 but it is reduced by 12% in a sale. Calculate the cost of the holiday in the sale.

5 There are 760 students at a school.

- **a** If 5% of the students are absent on one day, how many are at school?
- **b** If 15% of the students are in grade 8, how many is this?
- **c** If 45% of the students are boys, how many of the students are girls?
- **d** If the school population increases by 10%, how many students will it have?
- **e** If the school population decreases by 10%, how many students will it have?

6 In a written survey sent to 2740 people, only 1120 responded. What percentage (to one decimal place) responded to the survey?

7 Jane was earning K220 per week and then got a $7\frac{1}{2}$% pay rise. How much was she earning after the pay rise?

8 Isaac bought a bike for K145 and then sold it so that he made a 15% profit. At what price did he sell the bike?

9 After a pay increase of 5%, Lydia's salary was K18 165. What was her salary before the pay increase?

10 The price of a car was increased from K12 650 to K13 125. Calculate the percentage increase in the price to two decimal places.

11 The population of a town decreased from 1345 to 1258 over a two-year period. Calculate the percentage decrease in the population to two decimal places.

12 A sales person made K7380. On top of this, she was paid 5% commission for everything she earned over K2000.

- **a** How much money was she paid commission on?
- **b** How much commission was she paid?
- **c** How much was she paid in total?

13 A company made a profit of K37 000. If it decreases its profit by the following percentages, how much will it make?

a 10% b $7\frac{1}{2}$% c 5% d 15% e 12% f 8%

14 In a village, 55% of the people are female and the rest are male. If there are 88 females, how many males are there?

15 A holiday package costs K3619 including VAT of 10%.

a How much VAT has been included in the price?

b What is the cost of the holiday package before VAT is added?

16 A company employs 1540 people. Of these people, 897 have full-time jobs and the rest have part-time jobs.

a What percentage have full-time work?

b What percentage have part-time work?

17 Steven works for a fitness company that pays him a wage, plus 7% of the joining fee when he signs up new members. If the joining fee is K125, how much commission does Steven make if he signs up:

a 5 new members b 12 new members c 8 new members

d 20 new members e 15 new members f 18 new members?

18 At one store, a pair of trousers is advertised as $12\frac{1}{2}$% off the marked price of K80. At another store, a similar pair of trousers is advertised at 15% off the marked price of K85.

a At which store are the trousers cheaper?

b What is the difference in the discounted price between the two stores?

19 A group of schools raised money for charity. One school raised K218.75, which was 35% of the total amount raised by the schools. How much did the schools raise in total?

20 Alice bought her car for K11 780 and three years later, she sells it for K6 855. What is her percentage loss?

21 A tennis club has 150 members. The number of members increases by 20% each year. How many members does the club have after two years?

8.1.5 Apply ratios in solving problems from real life

Simplify ratios

Help Box

Ratios can be simplified to an equivalent ratio as long as the quantities or amounts are expressed in the same unit. A ratio must contain whole numbers.

Example: a ratio of 10 centimetres to 1 metre is the same as 10 centimetres to 100 centimetres and can be written as 10:100, which can be simplified to 1:10 by dividing both numbers by 10.

1 Write each of these ratios in its simplest form.

a	5:20	**b**	6:54	**c**	8:64	**d**	33:3	**e**	2:28	**f**	35:28
g	42:63	**h**	18:12	**i**	25:60	**j**	32:24	**k**	15:27	**l**	36:48
m	20:45	**n**	84:54	**o**	56:28	**p**	63:14	**q**	25:95	**r**	24:126

2 Convert the amounts in each ratio to the same unit before writing each in its simplest form.

a	3 cm:9 m	**b**	2 cm:3 mm	**c**	20t:K4
d	15 kg:800 g	**e**	K8:60t	**f**	20 minutes:3 hours
g	150 m:1 km	**h**	5 L:50 mL	**i**	6 hours:2 days
j	1 tonne:90 kg	**k**	15 seconds:$2\frac{1}{2}$ minutes	**l**	4 cm:5 m
m	240 mL:6 L	**n**	9 cm:8 mm	**o**	1 km:10 m
p	K2:80t	**q**	125 cm:3 m	**r**	750 mL:4 L

Help Box

The terms in a ratio must be whole numbers, so when comparing fractions or decimals we need to change them to whole numbers. This can be done in the following ways.

Example 1: $\frac{2}{5}:\frac{3}{10}$

Find the lowest common multiple of 5 and 10 (10).

Multiply both terms by 10. $\frac{2}{5} \times 10:\frac{3}{10} \times 10$

$= \frac{20}{5}:\frac{30}{10} = 4:3$

Example 2: $\frac{2}{5}:6$

Find the lowest common multiple of 5 and 1 (5).

Multiply both terms by 5. $\frac{2}{5} \times 5:\frac{6}{1} \times 5$

$= \frac{10}{5}:\frac{30}{1} = 2:30 = 1:15$

Example 3: 1.6:0.75

Make whole numbers by multiplying by 10, 100 or 1000. In this example multiply by 100.

1.6 × 100:0.75 × 100 = 160:75

Divide 160:75 by 5 to simplify to 32:15.

3 Write each of these as a correct ratio in its simplest form.

a	$\frac{3}{7}:\frac{5}{7}$	**b**	$\frac{2}{11}:\frac{4}{11}$	**c**	$\frac{3}{4}:\frac{1}{8}$	**d**	$\frac{2}{3}:\frac{5}{9}$	**e**	0.7:0.9	**f**	1.2:0.36
g	2.3:1.5	**h**	0.8:1.45	**i**	$\frac{3}{5}:\frac{7}{10}$	**j**	$\frac{4}{5}:\frac{3}{4}$	**k**	$\frac{1}{3}:\frac{2}{5}$	**l**	$\frac{7}{8}:\frac{1}{2}$

m 1.3:0.7 n 1.5:6 o 10:2.5 p 3.5:0.25 q $\frac{3}{5}$:6 r 8:$\frac{4}{10}$

s $\frac{3}{4}$:3 t 5:$4\frac{1}{5}$ u $\frac{4}{5}$:8 v 10:$\frac{2}{3}$ w $2\frac{1}{3}$:7 x 24:$1\frac{3}{5}$

Find equivalent ratios

Remember

Equivalent ratios can be made by multiplying or dividing all terms in the ratio by the same number.

1 Complete these equivalent ratios by multiplying.

a 1:3 = 4:$\square$ b 7:9 = $\square$:27 c 4:3 = 16:$\square$ d 2:3 = $\square$:15

e 6:5 = $\square$:25 f 3:5 = 9:$\square$ g 2:7 = $\square$:42 h 9:2 = 45:$\square$

i 1:8 = $\square$:56 j 3:2 = $\square$:24 k 7:4 = 21:$\square$ l 2:5 = 16:$\square$

m 9:4 = 18:$\square$ n 5:8 = $\square$:32 o 6:7 = 36:$\square$ p 11:7 = $\square$:28

2 Complete these equivalent ratios by dividing.

a 21:28 = $\square$:4 b 14:6 = 7:$\square$ c 40:15 = $\square$:3 d 10:16 = 5:$\square$

e 12:30 = 2:$\square$ f 24:60 = $\square$:5 g 27:18 = 3:$\square$ h 15:45 = $\square$:3

i 20:32 = $\square$:8 j 8:14 = 4:$\square$ k 54:63 = 6:$\square$ l 28:40 = $\square$:10

m 60:20 = $\square$:1 n 16:24 = 2:$\square$ o 32:18 = $\square$:9 p 38:24 = 19:$\square$

3 Complete these equivalent ratios by deciding whether to multiply or divide.

a 36:8 = $\square$:2 b 3:4 = 33:$\square$ c 9:8 = $\square$:48 d 16:36 = 4:$\square$

e 3:5 = 21:$\square$ f 9:12 = $\square$:4 g 10:25 = 2:$\square$ h 7:12 = $\square$:72

i 8:3 = $\square$:12 j 26:6 = 13:$\square$ k 14:8 = $\square$:4 l 18:33 = 6:$\square$

m 16:40 = 2:$\square$ n 5:13 = $\square$:52 o 3:2 = 45:$\square$ p 20:18 = $\square$:9

4 These equivalent ratios have three terms. Complete them by multiplying or dividing.

a 3:5:5 = $\square$:$\square$:30 b 2:5:7 = 12:$\square$:$\square$ c 4:8:6 = $\square$:4:$\square$

d 5:1:3 = 25:$\square$:$\square$ e 6:3:9 = $\square$:1:$\square$ f 8:5:3 = 24:$\square$:$\square$

g 12:8:12 = $\square$:$\square$:3 h 15:25:5 = 3:$\square$:$\square$ i 3:7:4 = $\square$:$\square$:20

j 3:5:7 = 12:$\square$:$\square$ k 18:30:24 = $\square$:5:$\square$ l 7:2:10 = $\square$:6:$\square$

m 16:10:14 = $\square$:$\square$:7 n 5:2:3 = 35:$\square$:$\square$ o 4:7:4 = $\square$:63:$\square$

Use ratio to share a quantity

Help Box

Example 1: *Share 40 into two parts in the ratio 5:3*

There are 8 parts in the ratio (5 + 3) so write each part as a fraction: $\frac{5}{8}$ and $\frac{3}{8}$

$\frac{5}{8}$ of 40 = 25 and $\frac{3}{8}$ of 40 = 15, so 40 shared in the ratio 5:3 is 25 and 15.

Example 2: *Share 50 into three parts in the ratio 3:5:2*

There are 10 parts in the ratio (3 + 5 + 2) so write each part as a fraction: $\frac{3}{10}$, $\frac{5}{10}$ and $\frac{2}{10}$

$\frac{3}{10}$ of 50 = 15, $\frac{5}{10}$ of 50 = 25 and $\frac{2}{10}$ of 50 = 10, so 50 shared in the ratio 3:5:2 is 15, 25 and 10.

1 Write each term in these ratios as a fraction of the whole.

a	2:3	**b**	4:5	**c**	1:3	**d**	5:2	**e**	7:2	**f**	1:6
g	3:8	**h**	4:1	**i**	4:2	**j**	5:7	**k**	3:5	**l**	8:2
m	2:7:1	**n**	6:5:3	**o**	4:3:7	**p**	5:1:8	**q**	1:4:2	**r**	3:4:4

2 Share K80 in the ratio:

a	1:3	**b**	5:3	**c**	3:2	**d**	3:7	**e**	9:1	**f**	1:4
g	1:7	**h**	7:9	**i**	17:3	**j**	5:11	**k**	11:9	**l**	13:27
m	5:8:3	**n**	9:1:10	**o**	2:3:5	**p**	7:5:8	**q**	4:1:3	**r**	19:8:13

3 Share 600 mL in the ratio:

a	1:5	**b**	5:7	**c**	3:7	**d**	1:2	**e**	3:5	**f**	8:7
g	11:4	**h**	16:9	**i**	19:1	**j**	17:13	**k**	5:19	**l**	23:17
m	4:7:4	**n**	6:1:3	**o**	1:4:3	**p**	7:5:8	**q**	4:3:5	**r**	7:4:9

4 Share 360 g in the ratio:

a	5:3	**b**	4:1	**c**	3:6	**d**	1:3	**e**	7:5	**f**	4:2
g	1:2	**h**	2:3	**i**	7:2	**j**	3:7	**k**	9:1	**l**	1:7
m	8:4:3	**n**	5:2:5	**o**	6:4:8	**p**	2:11:7	**q**	5:3:4	**r**	10:2:6

5 Share 240 cm in the ratio:

a	3:2	**b**	1:9	**c**	4:8	**d**	5:1	**e**	5:10	**f**	9:7
g	4:11	**h**	1:3	**i**	7:5	**j**	7:13	**k**	11:5	**l**	19:5
m	1:7:4	**n**	8:3:5	**o**	5:11:8	**p**	9:1:5	**q**	3:2:11	**r**	5:1:14

6 Share 1200 L in the ratio:

a	2:13	**b**	7:5	**c**	9:1	**d**	8:17	**e**	13:7	**f**	11:13
g	15:1	**h**	3:5	**i**	11:9	**j**	14:11	**k**	7:9	**l**	1:19
m	3:17:5	**n**	6:19:5	**o**	2:1:21	**p**	9:3:4	**q**	8:5:7	**r**	10:7:7

Solve ratio word problems

1 Concrete is made by mixing cement with sand and gravel in the ratio 2:5:5.

- **a** How many buckets of cement should be mixed with 30 buckets of gravel?
- **b** How many buckets of sand should be mixed with eight buckets of cement?

2 When mixing deep purple paint, the ratio of blue paint to red paint is 5:1.
If three small tins of red paint are used, how many small tins of blue paint are needed?

3 Nuts and dried fruits are mixed together in the ratio 4:3 to make a delicious treat.

- **a** If two cups of nuts are used, how many cups of dried fruits will be needed?
- **b** If $4\frac{1}{2}$ cups of dried fruit are used, how many cups of nuts will be needed?

4 A length of rope is cut into three pieces in the ratio 2:5:7. If the smallest piece is 10 metres long:

- **a** What is the length of each of the other pieces?
- **b** How long was the piece of rope before it was cut into three pieces?

5 At a school, the ratio of teachers to students is 1:28.

- **a** If there are nine teachers, how many students attend the school?
- **b** If there are 168 students, how many teachers are there?

6 A plaster contains a mix of water and powder in a ratio of 6:5. If 4.4 L of plaster has been mixed, what quantity of each ingredient has been used?

7 The ratio of concentrate to water when mixing a garden fertiliser is 1:80. If the following quantities of concentrate are used, how much water needs to be added?

a 5 mL **b** 20 mL **c** 12 mL **d** 8 mL

8 The ratio of water to chemical concentrate in a garden spray is 5:2.

- **a** If 500 mL of chemical concentrate is used, how much water must be added?
- **b** If 800 mL of water is used, how much chemical concentrate must be added?
- **c** If my container holds 8.4 litres, how much of each ingredient should I add to the container so that it is full of garden spray?

9 A recipe requires sugar, flour and cocoa powder in the ratio 1:4:2.

- **a** How many cups of flour should be mixed with one cup of cocoa powder?
- **b** How many cups of cocoa powder should be mixed with $1\frac{1}{2}$ cups of sugar?
- **c** How many cups of sugar should be mixed with 10 cups of flour?

10 If VAT is applied in the ratio 1:10, how much VAT is included in each of these prices?

a K187 **b** K137.50 **c** K253 **d** K104.50

11 A container holding 120 litres of water was shared between three smaller containers in the ratio 2:5:3. How much water did each of the smaller containers hold?

12 A city has a population of 8425.

a How many adults live in the city if the ratio of adults to children is 3:2?

b How many children live in the city?

13 At a school camp, the ratio of adults to students was 1:6.

a If there were eight adults, how many students were there?

b If there were 90 students, how many adults were there?

14 A model boat is built to a scale of 1:80. If the model boat is 40 cm long, how long is the actual boat in metres?

15 The ratio of flavouring to milk is 1:5.

a How much milk should be added to 50 mL of flavouring?

b How much milk should be added to 35 mL of flavouring?

c How much flavouring should be added to 2 litres of milk?

16 A hamburger recipe contains a mix of beef and pork mince in the ratio 3:1.

a If 720 g of beef mince is used, how much pork mince will be in the recipe?

b If 215 g of pork mince is used, how much beef mince will be in the recipe?

c If the total amount of mince used in the recipe is 1.38 kg, how much of each type of mince is used?

17 Table tops are made with the width and length in the ratio 2:5.

a If the width is one metre, what is the length?

b If the length is 1.25 metres, what is the width?

c If the width is 150 cm, what is the length?

18 A jar is filled with 10t and 20t coins in the ratio 4:3. There are thirty-six 20t coins in the jar.

a How many 10t coins are in the jar?

b How many coins are in the jar altogether?

19 There are 600 students at a school. The ratio of girls to boys is 3:5. How many girls are at this school?

20 The perimeter of a rectangle is 28 centimetres. The ratio of its length to its width is 5:2.

a What is the length of the rectangle?

b What is the width of the rectangle?

c What is the area of the rectangle?

8.1.6 Apply rates to solve simple problems from real life

8.3.5 Use time-rate calculations

Understand and solve rate problems

Remember

A rate compares quantities of a different kind and can only compare two quantities at a time.

1 What quantities are being compared in each of the following rates?

- **a** Peter earns K15 per hour.
- **b** The fabric costs K7 per metre.
- **c** A car travels at 70 km per hour.
- **d** Dog food costs K2.50 per 500 g.
- **e** A tap leaks water at 50 mL per minute.
- **f** An author writes 10 pages per week.
- **g** A bakery bakes 250 bread rolls per hour.
- **h** A tree grows 5 cm per week.
- **i** A gardener charges K8 per hour.
- **j** Sam walked 5 kilometres in one hour.
- **k** Joe's heart beats 72 times per minute.
- **l** One litre of paint covers 7 m^2.

2 Use multiplication to solve these rate problems.

- **a** At a fair, 120 drinks were sold on the first day. At this rate, how many drinks will be sold in one week?
- **b** Paint costs K8.50 per litre. What will 12 litres cost?
- **c** A running tap uses 15 litres of water per minute. How much water will it use in half an hour?
- **d** If Felix plants 42 seedlings per square metre, how many seedlings will be planted in 18 square metres?
- **e** If Lillian rides 12 kilometres per hour on her bike, how many kilometres will she ride in 2.5 hours?
- **f** It is recommended that lawn seed be planted at a rate of 45 grams per square metre. How many grams of lawn seed are required to plant 35 square metres?
- **g** If a secretary types 52 words per minute, how many will she type in 3.5 minutes?

3 Use division to solve these rate problems.

- **a** John worked for six hours and was paid K54. How much is this per hour?
- **b** A farmer spent K396 for 18 metres of fencing. How much is this per metre?
- **c** A hire car used 46 litres of petrol on a trip of 529 kilometres. How many kilometres did the car travel per litre of petrol?
- **d** Esta took 45 minutes to water her garden. The gauge on the tank showed that she had used 675 litres of water. How much water is this per minute?
- **e** A large family used 69 024 litres of water over 96 days. What was the average amount of water used per day?
- **f** A landscaping business delivered 2 tonnes of mulch in 25 days. What was the average number of kilograms delivered per day?
- **g** Maxine types at a rate of 64 words per minute. At this rate, how long would it take her to type a 1000-word essay?

Solve rate word problems

1 If an athlete ran 200 metres in 25 seconds, how many metres per second did he run?

2 How long will it take to travel 60 kilometres at the following speeds?

- **a** 10 km per hour
- **b** 12 km per hour
- **c** 15 km per hour

3 If Eileen earns K12.50 per hour, how long will it take her to earn K1000?

4 A car travels for 3.25 hours at a speed of 78 km per hour. How far, in kilometres, did the car travel?

5 Water is coming out of a tap at a rate of 96 litres per hour.

- **a** How many litres per minute is this?
- **b** How many millilitres per second is this?

6 Water is dripping from a tap at a rate of 64 drops every 8 minutes. How long will it take for the tap to drip 400 drops?

7 Alex leaves home and drives to his friend's village 108 kilometres away. How far from the village is he after driving for $1\frac{1}{4}$ hours at 60 km per hour?

8 What distance would a plane flying at 880 kilometres per hour travel in the following times?

a 3.5 hours **b** 2.75 hours **c** $5\frac{1}{2}$ hours **d** 45 minutes

9 A landowner has 15 square kilometres of land on which to graze cattle. If he wants to graze no more than 35 cattle per square kilometre, what is the maximum number of cattle that he can have?

10 What is the flying time of a plane that travels for 1610 kilometres at 460 km per hour?

11 A catering company charges K27.50 per person to provide and serve dinner. How much would the company charge for:

- **a** 35 people
- **b** 24 people
- **c** 16 people
- **d** 47 people
- **e** 85 people
- **f** 60 people?

12 A cook can make hamburgers at a rate of three every 5 minutes. If he works for 1.5 hours, how many hamburgers will he make?

Compare rates to find the best deal

1 Compare each pair of prices to find which one is better value.

- **a** 500 g of margarine at K5.50 or 750 g of margarine at K7.45
- **b** A 300 g jar of coffee at K15.60 or a 175 g jar of coffee at K9.30
- **c** 1 kg of soap powder at K8.45 or 1.5 kg of soap powder at K12.90
- **d** A 2 kg bag of rice for K5.25 or a 5 kg bag of rice for K11.45
- **e** 750 mL of vegetable oil at K6.90 or 1 L of vegetable oil at K8.15
- **f** 2.5 kg of potatoes at K8.65 or 1.5 kg of potatoes at K5.95

2 Different bike hire companies charge different rates depending on the number of hours for which a bike is rented. The chart below shows the one-off hire fee plus the hourly and daily rates charged by three companies. If a bike is hired for part of a day after the first or subsequent day, it is charged at the hourly rate.

	Hire fee	Rate per hour (8 hours or less)	Rate per day (more than 8 hours)
Company A	K3	K1.50	K11
Company B	K5	K1.20	K10.50
Company C	K3.50	K1.40	K10.25

Which company has the best deal in the following situations?

- **a** you want to hire a bike for 5 hours
- **b** you want to hire a bike for 8 hours
- **c** you want to hire a bike for 2 days
- **d** you want to hire a bike for 3 days and 5 hours
- **e** you want to hire a bike for 1 day and 2 hours

3 Calculate the best deal on the above situations if Company A reduced its hourly rate to K1.45.

4 This chart shows the cost of a single room per night at four hotels. The rate at each hotel varies according to the day of the week you wish to book your stay.

	Mon/Tues/Wed/Thurs	Friday	Sat/Sun
Hotel A	K95	K110	K115
Hotel B	K105	K108	K110
Hotel C	K92.50	K105	K120
Hotel D	K108	K115	K115

Calculate the best hotel deal for the following stays.

- **a** Friday, Saturday, Sunday and Monday nights
- **b** Saturday, Sunday and Monday nights
- **c** Monday, Tuesday, Wednesday and Thursday nights
- **d** Sunday, Monday and Tuesday nights
- **e** Seven nights from Monday to Sunday

5 Calculate the best deal for the above stays if Hotel D offered a 5% discount on each of its daily rates.

+20 +19 +18 +17 +16 +15 +14 +13 +12 +11 +10 +9 +8 +7 +6 +5 +4 +3 +2 +1 0 −1 −2 −3 −4 −5 −6 −7 −8 −9 −10 −11 −12 −13 −14 −15 −16 −17 −18 −19 −20

8.1.7 Apply directed numbers in problem solving

Compare and ordered directed numbers

Remember

A plus sign can be used to show positive numbers but is not necessary.
Examples: +7 means the same as 7; +8 + +5 means the same as 8 + 5.

1 Write < or > to make these statements true. Use the number line to help you.

a	+10 ☐ +15	**b**	−9 ☐ 2	**c**	−14 ☐ −19	**d**	16 ☐ +11
e	−13 ☐ −7	**f**	+3 ☐ 12	**g**	0 ☐ −7	**h**	+18 ☐ 27
i	−15 ☐ −18	**j**	−27 ☐ −22	**k**	30 ☐ −4	**l**	−23 ☐ 8
m	−6 ☐ 0	**n**	+21 ☐ +17	**o**	−4 ☐ −25	**p**	−12 ☐ −3
q	37 ☐ −9	**r**	19 ☐ +14	**s**	−39 ☐ −53	**t**	−1 ☐ 10

2 Write the number from each of these groups that has the greatest value.

a	27, +23, −17, −34	**b**	−35, +16, −41, +3	**c**	+31, −12, −48, 0
d	−4, 89, +72, −95	**e**	+36, −77, 54, +39	**f**	−82, −45, −63, −17
g	+11, 49, −4, 33	**h**	−81, −94, −52, −78	**i**	19, 0, −29, +12
j	−38, −61, +7, 2	**k**	+24, −57, +45, 36	**l**	−44, −50, −19, −37
m	+70, 77, −72, +74	**n**	−28, −18, 8, −3	**o**	+67, +98, 105, 93
p	−107, −89, −92, −101	**q**	−45, +6, −81, +32	**r**	−54, 19, −27, +14

3 Put these sets of numbers in order from lowest value to highest value.

a	+18, −36, −27, 11, −8, +25, 16, −4	**b**	−32, 17, −51, +36, 23, −7, −19, +21
c	−29, −18, −76, −32, −15, 18, −54, −3	**d**	35, +61, −23, 89, −11, +70, −33, +40
e	+21, −22, 28, +56, −42, 38, +52, −27	**f**	17, −14, −6, +60, 55, −56, −13, 45
g	77, −57, −50, 0, +9, −39, +44, −44	**h**	−64, +86, 81, −48, −53, 78, −73, +80
i	−90, −87, +80, −66, 77, +91, 88, −80	**j**	54, −54, 60, −60, +47, −47, −21, +12
k	+35, −38, 42, +41, −37, −30, 40, +30	**l**	−19, −17, +15, 23, −27, +16, −21, −18
m	−7, 9, −15, +11, +16, −14, 8, −10	**n**	+76, 63, −71, +70, −17, −70, +1, 0
o	+125, −105, 108, −119, +118, −112	**p**	−134, −143, 130, +139, −140, 137
q	216, −220, +214, 210, −219, −217	**r**	187, −178, +170, +189, −186, −175
s	−155, −160, +154, −158, 164, −161	**t**	−287, −285, 290, 278, +286, −280
u	+247, 242, −204, −240, 243, −200	**v**	+312, −321, −315, −310, 315, −312

Add and subtract directed numbers

Help Box

When adding and subtracting directed numbers, if two signs occur side by side without a number between them, then the following rules apply:

Two *like* signs make a POSITIVE or PLUS; for example, $3 - -2 = 5$ (the two negative signs mean plus). So, $+ +$ and $- -$ are the same as $+$.

Two *unlike* signs make a NEGATIVE or MINUS; for example, $9 - +7 = 2$ (one negative and one positive sign mean minus). So, $+ -$ and $- +$ are the same as $-$.

1 Use $=$ or $\neq$ to complete the following examples.

a	$+7 - -3 \square 10$	**b**	$-8 + -4 \square -12$	**c**	$9 - +5 \square 14$	**d**	$8 + +6 \square 14$
e	$-3 + +8 \square -5$	**f**	$-2 - -7 \square 5$	**g**	$10 + -5 \square 15$	**h**	$+7 - +7 \square 0$
i	$-4 + -3 \square -7$	**j**	$+2 - -8 \square -6$	**k**	$-5 + +9 \square +4$	**l**	$-4 + -6 \square -10$
m	$9 - -3 \square 6$	**n**	$12 - +4 \square 16$	**o**	$+8 - -7 \square 15$	**p**	$-3 + +1 \square -2$
q	$-6 - -5 \square -11$	**r**	$+9 + -2 \square 7$	**s**	$-11 + +4 \square +7$	**t**	$+6 - -0 \square 6$

2 Calculate answers to these additions and subtractions.

a	$-6 + +7$	**b**	$+9 - +5$	**c**	$-4 - -3$	**d**	$+8 - -9$
e	$17 + -8$	**f**	$-11 - +4$	**g**	$+10 + +6$	**h**	$-12 + -8$
i	$+15 + -7$	**j**	$20 - -7$	**k**	$-9 + +7$	**l**	$-14 - -3$
m	$-8 + +6$	**n**	$15 - -9$	**o**	$17 + -6$	**p**	$-16 - +10$
q	$21 + -9$	**r**	$-13 - +6$	**s**	$+18 - -12$	**t**	$17 - +8$

3 Calculate answers to these problems.

a	$+6 - -8 + +7$	**b**	$-12 - +5 - -4$	**c**	$9 + -4 - +8$	**d**	$-10 + +9 - -7$
e	$-8 - +5 + -2$	**f**	$17 + -8 - -7$	**g**	$-3 + +6 - +8$	**h**	$+11 - +5 + +7$
i	$3 + +9 - -1$	**j**	$+8 + -4 + -4$	**k**	$12 + +8 - +9$	**l**	$-5 + +14 - +6$
m	$+10 + -9 + +13$	**n**	$-6 - -7 - -6$	**o**	$-4 + +6 + +1$	**p**	$17 - -8 + +12$
q	$18 - +7 + -9$	**r**	$+16 - -3 + -8$	**s**	$-13 - +10 + -11$	**t**	$-9 + +6 - -0$

4 Complete each number sentence by inserting a positive or negative number.

a	$8 - \square = 10$	**b**	$-4 + \square = -9$	**c**	$+6 - \square = 11$	**d**	$2 - \square = -5$
e	$\square + -4 = -7$	**f**	$\square - -8 = 2$	**g**	$\square - -3 = 13$	**h**	$\square - +9 = 7$
i	$-2 - \square = 6$	**j**	$14 + \square = -3$	**k**	$-10 - \square = -9$	**l**	$11 + \square = 0$
m	$\square - -10 = 22$	**n**	$\square - +7 = 17$	**o**	$\square + -6 = 24$	**p**	$\square + +11 = 8$
q	$19 + \square = 11$	**r**	$-17 - \square = -7$	**s**	$\square + -12 = -6$	**t**	$\square - -16 = 18$

Multiply and divide directed numbers

Help Box

When multiplying and dividing directed numbers, multiply and divide the two numbers in the usual way and then look at the signs to work out the final answer.

Two *like* signs make a POSITIVE or PLUS answer; for example, −8 × −3 = 24, −15 ÷ −3 = 5

So, + × +, − × −, + ÷ + and − ÷ − make a plus (+) answer.

Two *unlike* signs make a NEGATIVE or MINUS answer; for example, −3 × +4 = −12, 16 ÷ −2 = −8

So, + × −, − × +, + ÷ − and − ÷ + make a minus (−) answer.

1 Copy and complete these multiplication squares.

a

×	−3	5	−2	−7	4	11
−3						
6						
−5						
−4						
8						

b

×	6	−4	2	−8	−5	9
−2						
−9						
10						
7						
−12						

2 Solve these divisions.

a −16 ÷ 8 **b** 40 ÷ −8 **c** −16 ÷ −4 **d** +18 ÷ 3
e 21 ÷ −7 **f** −27 ÷ −3 **g** 25 ÷ −5 **h** −32 ÷ +4
i +15 ÷ −5 **j** 12 ÷ −3 **k** −30 ÷ +6 **l** +28 ÷ +4
m −56 ÷ −8 **n** 64 ÷ +8 **o** +24 ÷ −3 **p** −36 ÷ +9
q +20 ÷ −4 **r** −35 ÷ −7 **s** 42 ÷ −6 **t** +22 ÷ 11

Help Box

When multiplying more than two directed numbers together, do it in two steps.

For example, −3 × +5 × −4

Step 1: −3 × +5 = −15 Step 2: −15 × −4 = 60

So, −3 × +5 × −4 = 60.

3 Solve these multiplications.

a +3 × −6 × −2 **b** −4 × −3 × 8 **c** −6 × +4 × −5 **d** 7 × −5 × −2
e 5 × +8 × −3 **f** −6 × 6 × −3 **g** −7 × −3 × −4 **h** 2 × −9 × +4
i 10 × −8 × 2 **j** −5 × −5 × −5 **k** +4 × 3 × −6 **l** −2 × −11 × 2
m 9 × −9 × −3 **n** −1 × −15 × 4 **o** 8 × +6 × −1 **p** −5 × 10 × 7
q +6 × 0 × −8 **r** −11 × +3 × −5 **s** 4 × −12 × −4 **t** −7 × −2 × −8

Add, subtract, multiply and divide directed numbers

Help Box

The rules for the order of operations also apply to directed numbers. The rules are:

1. brackets first
2. then multiplication and division from left to right
3. finally, addition and subtraction from left to right

Example: −5 × (8 + −3) − 5
Brackets first: −5 × (5) − 5
Then × or ÷: −25 − 5
Then + or −: −30
So, −5 × (8 + −3) − 5 = −30.

1 Use the correct order of operations to solve these problems.

a −16 + −12 ÷ −2
b −40 ÷ −10 + 6
c −3 + (−36 ÷ 6) × 2
d 18 − 6 + −2 × 8
e (−5 + −8) × 3 + 12
f (32 + 8) ÷ (−4 × −5)
g 4 × 6 + −5 × 3
h −7 × (8 + −3) + 25
i 27 ÷ −9 × 2 + −6 × 5
j −10 × (8 + −2) − 3
k −4 + 20 ÷ −5 − (−3 + −7)
l 16 ÷ −8 + −3 × 3
m (20 + −4) ÷ (−2 × −4)
n −9 × 5 + −11
o −50 ÷ −10 + −3

2 Copy and complete these addition squares.

a

+	−12	8	−6	+15	−22	−9
−7						
24						
−8						
+19						
−30						
45						

b

+	18	−7	−14	11	−20	16
+25						
−12						
−17						
50						
−29						
−15						

3 Copy and complete these division circles.

a

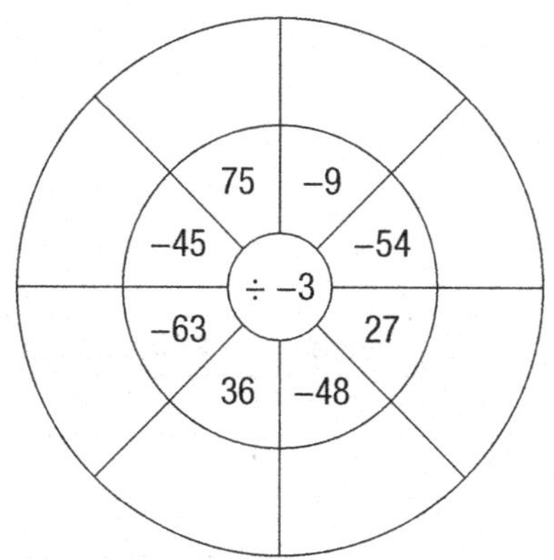

b

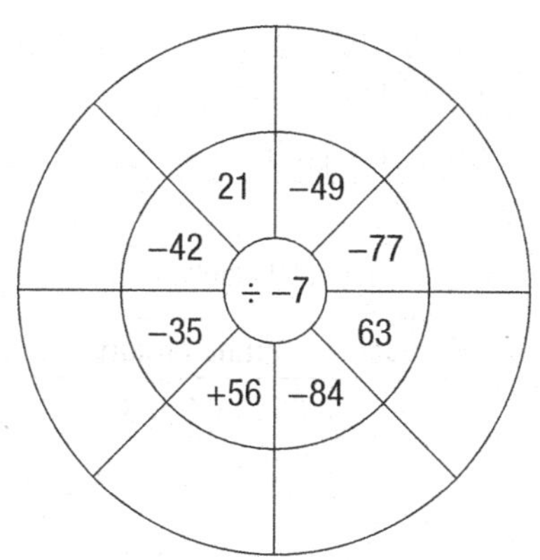

8.1.8 Use integer indices and fractional indices where the answers are rational

Find prime factors with factor trees

Help Box

A factor tree finds two factors at a time until all the factors are prime numbers. The original number can then be expressed as a product of its prime factors. For example, the number 81 can be written as $3 \times 3 \times 3 \times 3$ or in index form as 3^4.

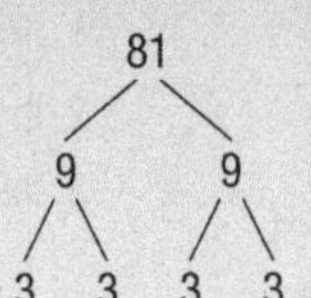

1 Copy and complete these factor trees to find the prime factors of each number and then express each in index form.

a

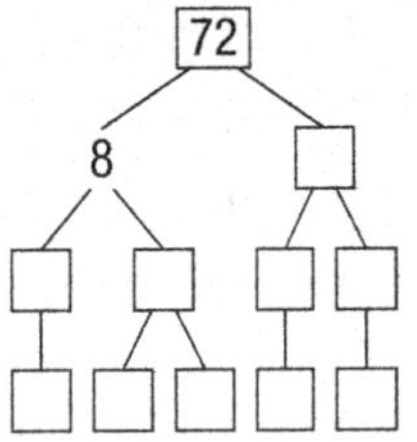

b

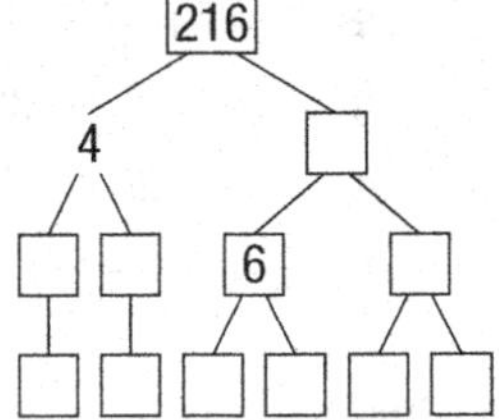

c

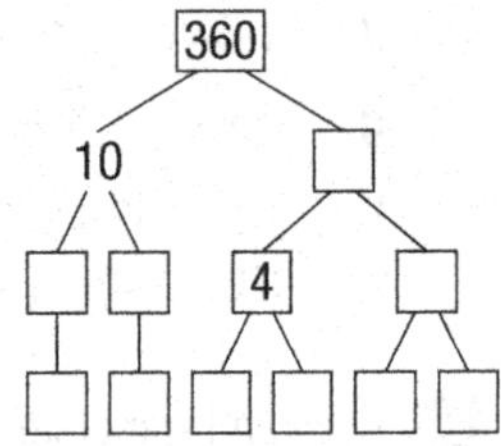

d

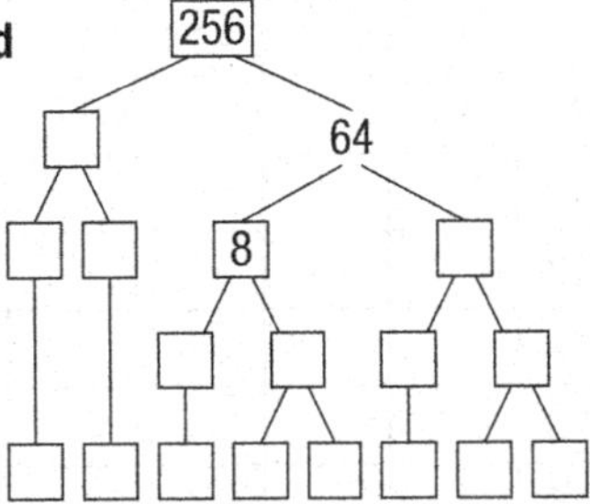

e

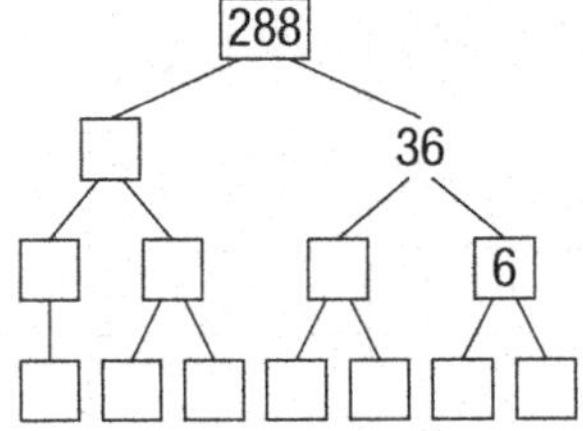

f

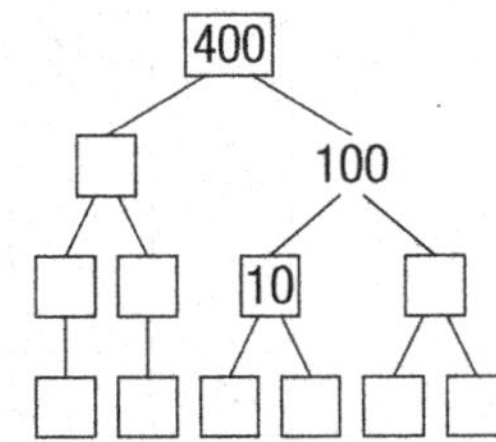

g

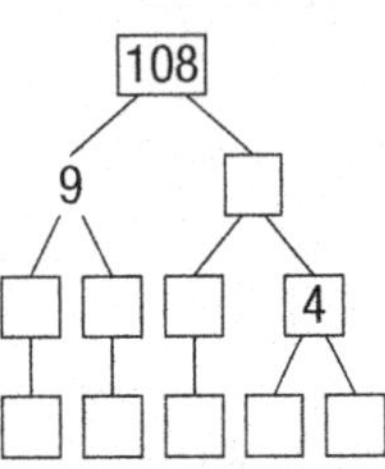

h

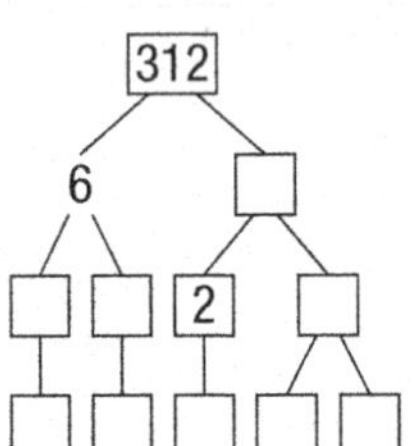

i

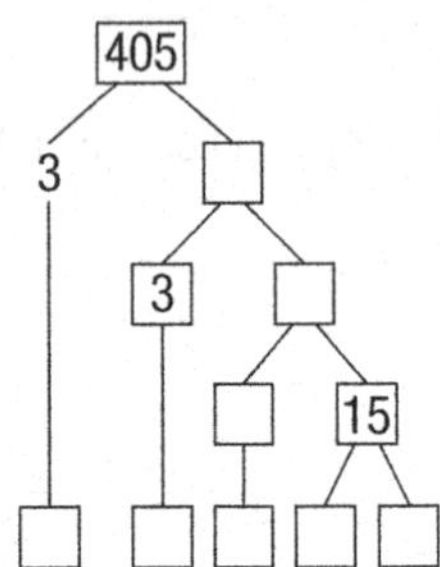

2 Copy and complete this table.

	Number	Prime factors	Index form
a		$2 \times 2 \times 2 \times 3 \times 3$	$2^3 \times 3^4$
b		$2 \times 2 \times 4 \times 4$	
c	96		
d			$2^4 \times 3^3$

	Number	Prime factors	Index form
e	441		
f			$2^3 \times 5^2$
g		$2\times2\times2\times2\times7\times7$	
h	600		

Understand squares, cubes, square roots and cube roots

Help Box

The opposite of a square number is a square root.
The symbol $\sqrt{}$ means square root.
Example: the $\sqrt{36} = 6$ because 6×6 (6^2) $= 36$

The opposite of a cube number is a cube root.
The symbol $\sqrt[3]{}$ means cube root.
Example: the $\sqrt[3]{8} = 2$ because $2 \times 2 \times 2$ (2^3) $= 8$

Remember

Any number that has an index or power of 2 can be referred to as squared. Any number that has an index or power of 3 can be referred to as cubed.

1 Calculate these squared and cubed numbers.

a	5^2	**b**	4^3	**c**	11^2	**d**	2^3	**e**	7^2	**f**	3^3
g	10^3	**h**	15^2	**i**	5^3	**j**	20^2	**k**	9^3	**l**	12^2
m	$4^2 + 8^2$	**n**	$6^3 - 9^2$	**o**	$5^3 + 5^2$	**p**	$8^3 - 15^2$	**q**	$11^3 + 7^2$	**r**	$20^2 - 7^3$
s	$2^3 \times 3^2$	**t**	$8^2 \times 10^3$	**u**	$8^2 \div 2^3$	**v**	$11^2 \times 3^3$	**w**	$4^3 \times 2^2$	**x**	$5^3 \div 5^2$

2 Find the square root of these numbers.

a	25	**b**	100	**c**	256	**d**	121	**e**	625	**f**	49
g	324	**h**	900	**i**	196	**j**	81	**k**	144	**l**	1024
m	36	**n**	9	**o**	225	**p**	289	**q**	400	**r**	784

3 Find the cube root of these numbers.

a	27	**b**	1000	**c**	125	**d**	1331	**e**	8000	**f**	64
g	343	**h**	729	**i**	216	**j**	1728	**k**	3375	**l**	512
m	8	**n**	2197	**o**	4096	**p**	2744	**q**	6859	**r**	4913

Remember

A positive number and a negative number squared will both equal the same positive number; for example, 5^2 and -5^2 both equal 25. Therefore, the $\sqrt{25} = 5$ and -5.

A positive number and a negative number cubed will have different answers; for example, $2^3 = 8$, whereas $-2^3 = -8$. Therefore, a number can only have one cube root.

4 Which is the larger number in each pair?

a $\sqrt{36}$ or 2^3
b -4^2 or $\sqrt[3]{125}$
c $\sqrt{100}$ or $\sqrt[3]{1331}$
d -5^3 or $\sqrt[3]{1000}$
e 7^2 or -7^3
f $\sqrt{64}$ or 9^2
g -9^2 or 4^3
h $\sqrt{64}$ or -2^3
i 3^3 or $\sqrt{400}$
j $\sqrt{16}$ or $\sqrt[3]{27}$
k $\sqrt{144}$ or $\sqrt[3]{1000}$
l -3^3 or 3^3
m $\sqrt{121}$ or 4^2
n 8^3 or $\sqrt[3]{8000}$
o -7^3 or $\sqrt{49}$
p $\sqrt{225}$ or 4^2

5 Use >, < or = to complete these.

a -4^2 ☐ $\sqrt{256}$
b $\sqrt{81}$ ☐ 2^3
c 7^2 ☐ -7^2
d $\sqrt[3]{1000}$ ☐ $\sqrt{100}$
e -5^3 ☐ 10^2
f -2^3 ☐ $\sqrt{64}$
g $\sqrt{144}$ ☐ $\sqrt[3]{216}$
h 3^3 ☐ $\sqrt{400}$
i $\sqrt{121}$ ☐ -3^2
j $\sqrt[3]{512}$ ☐ 2^3
k 20^2 ☐ 7^3
l $\sqrt{225}$ ☐ -4^2
m 9^3 ☐ -25^2
n 5^2 ☐ $\sqrt{900}$
o $\sqrt[3]{343}$ ☐ $\sqrt{49}$
p -4^3 ☐ $\sqrt{64}$

Explore positive and zero indices

Help Box

Index numbers (indices) are used to show that a number is multiplied by itself a certain number of times.
Example: $3^4 = 3 \times 3 \times 3 \times 3$ so 4 is the index number that shows 3 is multiplied by itself 4 times.
An index of 1 gives the same result as the number itself; for example, 3^1 means 3, 5^1 means 5 and so on.
An index of 0 always gives a result of 1; for example, 2^0 means 1, 7^0 means 1 and so on.

1 Write each of these in index notation.

a 2 to the power of 7
b 8 to the fifth power
c 4 squared
d 11 cubed
e 6 to the power of 4
f five 8s multiplied together
g 5 to the seventh power
h 15 squared
i 3 to the power of 9

2 Copy and complete this table. You can use a calculator to help you.

$2^0 = 1$	$2^1 = 2$	$2^2 = 4$	$2^3 = 8$	$2^4 =$	$2^5 =$
$3^0 =$	$3^1 =$	$3^2 =$	$3^3 =$	$3^4 =$	$3^5 =$
$4^0 =$	$4^1 =$	$4^2 =$	$4^3 =$	$4^4 =$	$4^5 =$
$5^0 =$	$5^1 =$	$5^2 =$	$5^3 =$	$5^4 =$	$5^5 =$
$6^0 =$	$6^1 =$	$6^2 =$	$6^3 =$	$6^4 =$	$6^5 =$
$7^0 =$	$7^1 =$	$7^2 =$	$7^3 =$	$7^4 =$	$7^5 =$
$8^0 =$	$8^1 =$	$8^2 =$	$8^3 =$	$8^4 =$	$8^5 =$
$9^0 =$	$9^1 =$	$9^2 =$	$9^3 =$	$9^4 =$	$9^5 =$
$10^0 =$	$10^1 =$	$10^2 =$	$10^3 =$	$10^4 =$	$10^5 =$

3 Fill the gap with >, < or =.

a	$6^0 \square 12^0$	b	$7^1 \square 4^1$	c	$2^1 \square 1^2$	d	$9^0 \square 9^1$
e	$8^1 \square 2^3$	f	$10^2 \square 100^0$	g	$25^1 \square 5^2$	h	$4^2 \square 12^1$
i	$3^5 \square 5^3$	j	$2^8 \square 8^2$	k	$9^4 \square 4^9$	l	$125^1 \square 5^3$
m	$2^5 \square 5^2$	n	$-7^2 \square 49^1$	o	$8^2 \square -4^3$	p	$15^0 \square -2^4$

4 Solve these additions and subtractions.

a	$8^0 + 5^0 + 10^1$	b	$8^1 - 12^0$	c	$-2^1 + 2^2$	d	$10^2 - 10^1$
e	$6^1 + 4^1 + -4^1$	f	$7^2 - -7^2$	g	$5^1 + 9^0 + -3^1$	h	$3^3 - 25^1$
i	$-2^2 + 15^1 + 7^1$	j	$15^0 - 10^0$	k	$-8^1 + 6^0 + 1^1$	l	$45^1 - 23^0$
m	$8^3 + 8^0 - 13^1$	n	$12^1 - 2^3$	o	$135^0 + 275^0$	p	$16^1 - 4^2$

5 Solve these multiplications and divisions.

a	$8^0 \times 5^1$	b	$12^1 \times 3^1$	c	$7^2 \times 7^0$	d	$4^1 \times 2^3$	e	$10^0 \times 8^2$	f	$12^0 \times 6^0$
g	$5^2 \div 5^1$	h	$3^3 \div 3^0$	i	$8^1 \div 2^2$	j	$10^3 \div 10^1$	k	$30^1 \div 6^1$	l	$2^6 \div 8^1$

Explore negative indices

Help Box

As well as positive and zero index numbers (indices), there are also negative index numbers.

While positive index numbers use multiplication, negative index numbers use division.

A negative index shows how many times to divide by the number.

Examples: $8^{-1} = 1 \div 8 = \frac{1}{8}$ or 0.125; $10^{-2} = 1 \div 10 \div 10 = \frac{1}{100}$ or 0.01

1 Complete these equations.

a $5^{-1} = 1 \div 5 = \frac{1}{5}$ or _____

b $5^{-2} = 1 \div 5 \div 5 = \frac{1}{25} =$ _____

c $2^{-2} = 1 \div 2 \div 2 =$ _____ or _____

d $2^{-3} = 1 \div$ __ $\div$ __ $\div$ __ $=$ _____ or _____

e $4^{-2} = 1 \div$ __ $\div$ __ $=$ _____ or _____

f $10^{-1} = 1 \div$ __ $=$ _____ or _____

g $10^{-2} = 1 \div$ __ $\div$ __ $=$ _____ or _____

h $10^{-3} = 1 \div$ __ $\div$ __ $\div$ __ $=$ _____ or _____

i $3^{-1} = 1 \div$ __ $=$ _____ or _____

j $3^{-2} = 1 \div$ __ $\div$ __ $=$ _____ or _____

k $8^{-2} = 1 \div$ __ $\div$ __ $=$ _____ or _____

l $16^{-1} = 1 \div$ __ $=$ _____ or _____

2 Calculate the value of these negative indices (to three decimal places).

a	5^{-3}	b	9^{-1}	c	3^{-3}	d	12^{-1}	e	15^{-2}	f	11^{-2}
g	4^{-4}	h	3^{-4}	i	20^{-1}	j	2^{-4}	k	18^{-1}	l	2^{-6}

Help Box

Negative index numbers (indices) can also be calculated by changing them to a positive fraction.

Example 1: 10^{-2} means 1 divided by 10 two successive times:

$1 \div (10 \times 10) = 1 \div 100$

$= \frac{1}{100}$ or 0.01

Therefore, 10^{-2} is the same as $\frac{1}{10^2}$.

Example 2: 2^{-3} means 1 divided by 2 three successive times:

$1 \div (2 \times 2 \times 2) = 1 \div 8$

$= \frac{1}{8}$ or 0.125

Therefore, 2^{-3} is the same as $\frac{1}{2^3}$.

3 Write these negative indices as positive fractions.

a	5^{-2}	b	4^{-3}	c	7^{-2}	d	10^{-4}	e	3^{-5}	f	2^{-8}
g	8^{-1}	h	9^{-4}	i	6^{-2}	j	5^{-5}	k	11^{-2}	l	7^{-3}
m	12^{-4}	n	3^{-10}	o	8^{-7}	p	4^{-5}	q	10^{-9}	r	15^{-4}

4 Use the method shown above to calculate the value of these negative indices (to three decimal places).

a	10^{-3}	b	4^{-2}	c	6^{-1}	d	2^{-4}	e	3^{-3}	f	8^{-2}
g	12^{-2}	h	5^{-2}	i	2^{-5}	j	4^{-3}	k	7^{-3}	l	11^{-2}

5 Re-write these to their equivalent using a negative index number.

a	$\frac{1}{7^5}$	b	$\frac{1}{12^4}$	c	$\frac{1}{9^5}$	d	$\frac{1}{8^7}$	e	$\frac{1}{11^4}$	f	$\frac{1}{5^6}$
g	$\frac{1}{3^8}$	h	$\frac{1}{2^{10}}$	i	$\frac{1}{12^5}$	j	$\frac{1}{4^4}$	k	$\frac{1}{6^3}$	l	$\frac{1}{10^7}$
m	$\frac{1}{5^9}$	n	$\frac{1}{1^5}$	o	$\frac{1}{11^8}$	p	$\frac{1}{9^4}$	q	$\frac{1}{15^6}$	r	$\frac{1}{7^{10}}$

Assessment Number and Application

Fractions

1 Insert the symbol >, < or = to make a true statement.

a $\frac{8}{9} \square 1\frac{1}{8}$ b $\frac{9}{7} \square \frac{7}{9}$ c $1\frac{2}{3} \square \frac{5}{3}$ d $\frac{5}{6} \square \frac{5}{8}$

e $2\frac{3}{4} \square \frac{11}{4}$ f $\frac{3}{8} \square \frac{4}{7}$ g $\frac{11}{8} \square 1\frac{1}{4}$ h $\frac{7}{12} \square \frac{3}{5}$

2 Put each group of fractions and mixed numbers in order from smallest to largest.

a $1\frac{7}{8}$ $1\frac{3}{4}$ $\frac{17}{8}$ $\frac{8}{8}$ $\frac{3}{4}$ b $\frac{2}{3}$ $\frac{3}{2}$ $\frac{5}{6}$ $\frac{1}{6}$ $1\frac{1}{3}$ c $\frac{4}{5}$ $1\frac{2}{5}$ $\frac{1}{2}$ $\frac{9}{5}$ $\frac{3}{10}$

d $\frac{12}{7}$ $\frac{7}{12}$ $\frac{1}{4}$ $1\frac{1}{12}$ $\frac{2}{3}$ e $2\frac{1}{2}$ $\frac{5}{6}$ $\frac{11}{6}$ $\frac{1}{3}$ $\frac{6}{5}$ f $1\frac{2}{3}$ $\frac{7}{9}$ $\frac{18}{9}$ $\frac{2}{3}$ $\frac{1}{9}$

3 Add or subtract these fractions in your head. Write the answers in their simplest form.

a $1\frac{1}{8} + \frac{3}{4}$ b $4\frac{1}{4} - 2\frac{1}{2}$ c $2\frac{3}{10} + \frac{4}{5}$ d $1\frac{1}{2} - \frac{5}{6}$

4 Solve these fraction additions and subtractions and write the answers in their simplest form.

a $2\frac{3}{7} + 1\frac{2}{3}$ b $1\frac{7}{8} + 2\frac{3}{5}$ c $\frac{9}{10} + 3\frac{3}{4}$ d $3\frac{5}{8} + 2\frac{5}{6}$

e $\frac{7}{12} + 4\frac{1}{3}$ f $5\frac{8}{9} + 2\frac{1}{2}$ g $4\frac{1}{4} - 2\frac{1}{6}$ h $3\frac{2}{7} - 1\frac{2}{3}$

i $5\frac{3}{8} - 2\frac{4}{5}$ j $1\frac{5}{6} - \frac{7}{8}$ k $4\frac{3}{10} - 2\frac{3}{4}$ l $3\frac{2}{9} - \frac{5}{12}$

5 Solve these fraction multiplications and divisions and write the answers in their simplest form.

a $1\frac{4}{5} \times 1\frac{1}{3}$ b $2\frac{3}{4} \times \frac{5}{7}$ c $4\frac{1}{2} \times 1\frac{4}{7}$ d $\frac{7}{12} \times 2\frac{3}{8}$

e $2\frac{5}{6} \times 12$ f $\frac{7}{9} \times 4\frac{1}{4}$ g $3\frac{2}{3} \div 1\frac{1}{8}$ h $2\frac{5}{9} \div \frac{4}{5}$

i $3\frac{9}{10} \div 4\frac{2}{3}$ j $5\frac{3}{7} \div 8$ k $15 \div 3\frac{1}{6}$ l $8\frac{3}{4} \div 2\frac{7}{9}$

6 Solve these problems by first converting all the decimals to common fractions.

a $5.3 + 2\frac{2}{5}$ b $4\frac{3}{4} - 1.85$ c $2.23 \times 1\frac{2}{3}$ d $3\frac{3}{8} \div 0.7$

e $2\frac{1}{6} \times 1.47$ f $5.21 \div \frac{6}{7}$ g $1\frac{4}{9} + 3.07$ h $6.04 - 2\frac{1}{3}$

7 Solve these fraction word problems.

a A large container holds $23\frac{4}{5}$ kg of flour. How many smaller containers, each holding $2\frac{1}{4}$ kg of flour, can be filled from the large container?

b Each sack of potatoes weighs $7\frac{2}{3}$ kg. How much will 15 sacks of potatoes weigh?

c Naomi has a roll of fabric that is $4\frac{1}{8}$ metres long. If she uses $2\frac{3}{7}$ metres to make a dress, how much fabric is left?

Assessment Number and Application

Decimals

1 Insert the symbol < or > to make a true statement.

a 3.007 □ 30.77 **b** 65.06 □ 65.006 **c** 417.4 □ 417.42 **d** 8.59 □ 8.6

e 76.72 □ 76.7 **f** 34.045 □ 34.05 **g** 289.16 □ 289.2 **h** 50.08 □ 50.1

2 Copy and complete each table.

a

+	9.56	28.45	33.9	164.7
73.57				
5.686				
35.08				
281.4				

b

−	64.72	8.055	216.6	48.28
67.9				
263.94				
77.58				
4.201				

3 Solve these decimal multiplications.

a 63.8 × 100 **b** 9.044 × 10 **c** 36.28 × 70 **d** 512.6 × 90

e 492.3 × 6 **f** 7.047 × 23 **g** 59.65 × 7.4 **h** 135.8 × 4.9

i 246.7 × 2.18 **j** 74.06 × 32.4 **k** 189.45 × 6.7 **l** 48.25 × 3.06

4 Solve these decimal divisions.

a 83.27 ÷ 10 **b** 483.35 ÷ 100 **c** 662.4 ÷ 40 **d** 128.46 ÷ 60

e 723.1 ÷ 7 **f** 536.64 ÷ 16 **g** 593.4 ÷ 23 **h** 707.18 ÷ 19

i 324.52 ÷ 7.6 **j** 313.96 ÷ 4.7 **k** 255.42 ÷ 5.4 **l** 198.8 ÷ 3.5

5 Solve these decimal word problems.

a The capacity of a water tank is 1250.5 litres. Currently it contains 786.75 litres. How much more water is needed to fill the tank?

b If I buy 3.4 kg of sausages at K5.85 per kg and 2.7 kg of minced beef at K9.35 per kilogram, what is the total cost?

c The total cost of 56 books was K372.40. How much did each book cost if they were the same price?

d A 28-metre length of rope is cut into three pieces. One piece is 9.56 m and another piece is 10.78 m. Find the length of the third piece.

e A car has travelled 135.7 km out of a total distance of 334.25 km. How much further does the car have to travel to reach its destination?

Assessment — Number and Application

Convert between fractions, decimals, percentages and ratios

1 Write these decimals as common fractions or mixed numbers in their simplest form.

a 3.9 b 0.67 c 2.8 d 5.35 e 0.06 f 1.072
g 0.266 h 8.38 i 11.62 j 0.455 k 6.54 l 9.618

2 Write these common fractions and mixed numbers as decimals.

a $\frac{7}{100}$ b $1\frac{3}{5}$ c $\frac{47}{100}$ d $\frac{9}{50}$ e $5\frac{17}{25}$ f $3\frac{9}{20}$
g $8\frac{3}{1000}$ h $\frac{219}{500}$ i $\frac{63}{125}$ j $12\frac{18}{30}$ k $6\frac{77}{200}$ l $\frac{208}{250}$

3 Convert each of these fractions to decimals (to three places unless recurring) by dividing the numerator by the denominator.

a $\frac{8}{9}$ b $\frac{5}{8}$ c $\frac{5}{7}$ d $\frac{2}{11}$ e $\frac{11}{12}$ f $\frac{8}{15}$

4 Change these fractions and decimals to percentages.

a $\frac{43}{100}$ b $\frac{17}{20}$ c $2\frac{4}{5}$ d $8\frac{3}{10}$ e $\frac{7}{8}$ f $1\frac{4}{25}$
g 0.05 h 3.87 i 0.19 j 2.008 k 7.3 l 0.2

5 Change these percentages to fractions and decimals. Write each fraction in its simplest form.

a 58% b 3% c 89% d 45% e 175% f 230%
g 18% h 151% i 24% j $12\frac{1}{2}\%$ k $5\frac{1}{4}\%$ l $27\frac{1}{2}\%$

6 Change these ratios to fractions or mixed numbers reduced to their simplest form.

a 8:10 b 9:15 c 5:8 d 9:4 e 7:6 f 4:20
g 11:8 h 6:42 i 12:60 j 8:3 k 3:24 l 16:10

7 Write these fractions and mixed numbers as ratios in their simplest form.

a $\frac{9}{10}$ b $\frac{10}{12}$ c $\frac{4}{20}$ d $\frac{18}{33}$ e $\frac{12}{15}$ f $\frac{5}{30}$
g $4\frac{2}{3}$ h $\frac{11}{6}$ i $2\frac{4}{5}$ j $5\frac{2}{10}$ k $\frac{18}{11}$ l $1\frac{7}{8}$

8 Write each group in order from largest to smallest.

a 55%, 5:11, $\frac{7}{10}$, 0.67 b $\frac{3}{4}$, 0.075, 80%, 1:4 c 0.23, $\frac{2}{3}$, 2:4, 20%
d 11:4, 1.9, $2\frac{1}{4}$, 200% e 0.33, 35%, 3:5, $\frac{1}{5}$ f 10%, 1.0, 10:1, $\frac{5}{10}$

Assessment — Number and Application

Percentages

1 Write these numbers as percentages. Round your answers to one decimal place.

a 38 out of 160 b 15 out of 120 c 63 out of 195
d 675 out of 1560 e 812 out of 1346 f 1468 out of 3475

2 Express the first amount as a percentage of the second. Round your answers to one decimal place.

a 850 g out of 1 kg b 46 cm out of 2.5 m c 42t out of K5
d 345 mL out of 1.7 L e 685 kg out of 3.4 t f 518 m out of 1 km

3 Find these percentage increases.

a 40 by 25% b 34 by 10% c 58 by 15% d 29 by 5%
e 124 by 12% f 180 by 3% g 132 by $7\frac{1}{2}$% h 210 by 11.7%

4 Find these percentage decreases.

a 90 by 20% b 66 by 15% c 72 by 10% d 83 by 25%
e 156 by 8% f 214 by 16% g 185 by 1% h 355 by 30%

5 Calculate the percentage increase when:

a 45 becomes 80 b 34 becomes 51 c 29 becomes 42
d 87 becomes 112 e 109 becomes 130 f 164 becomes 180

6 Calculate the percentage decrease when:

a 52 becomes 30 b 75 becomes 58 c 94 becomes 67
d 106 becomes 87 e 153 becomes 121 f 187 becomes 162

7 Calculate the price after these discounts are given.

a 15% off K60 b 10% off K108.50 c 5% off K78
d $7\frac{1}{2}$% off K38 e 20% off K176 f $12\frac{1}{2}$% off K134

8 The following amounts of money represent 85% of the full amount. Calculate 100% of each amount.

a K65 b K44 c K90 d K230 e K182 f K325

9 Solve these percentage word problems.

a A 10-day trek costs K1255 including VAT of 10%. What is the cost of the trek before VAT is added?

b A sports club has 420 members. If the number of members increases by 15%, how many members would the club have?

c A water tank holds 2500 litres when full. When it is 55% full, how much water is needed to fill the tank?

Assessment — Number and Application

Multiple choice test for ratios and rates

1 Which ratio is equivalent to 18:27?

a 27:18 **b** 2:3
c 6:7 **d** 3:4

2 Which ratio is equivalent to 10t to K3?

a 10:3 **b** 10:30
c 1:30 **d** 1:3

3 Which is the correct ratio for $\frac{3}{8}$ to $\frac{5}{8}$?

a 3:5 **b** 3:8 **c** 5:8 **d** 8:8

4 Which is the correct ratio for $\frac{2}{3}$ to $\frac{5}{6}$?

a 2:3 **b** 4:5 **c** 2:5 **d** 3:6

5 Which is the correct ratio for 1.4 to 0.7?

a 2:1 **b** 4:7 **c** 1:2 **d** 1:7

6 Which ratio is equivalent to 4:7:5?

a 8:14:5 **b** 2:3:2
c 5:4:7 **d** 8:14:10

7 Which ratio is equivalent to 16:24:12?

a 1:2:1 **b** 4:6:3
c 4:8:3 **d** 8:10:6

8 How many parts are in the ratio 5:7?

a 57 **b** 7 **c** 12 **d** 2

9 How many parts are in the ratio 2:5:4?

a 11 **b** 254 **c** 9 **d** 7

10 What is 45 shared into two parts in the ratio 5:4?

a 25 and 20 **b** 40 and 5
c 20 and 25 **d** 30 and 15

11 What is 160 shared into two parts in the ratio 1:7?

a 10 and 70 **b** 140 and 20
c 10 and 150 **d** 20 and 140

12 What is 35 shared into three parts in the ratio 1:4:2?

a 10, 5, 10 **b** 10, 40, 20
c 5, 20, 10 **d** 10, 20, 30

13 Which ratio is *not* equivalent to 2:5?

a 4:10 **b** 16:40
c 6:12 **d** 20:50

14 Which ratio is *not* equivalent to 7:3?

a 21:9 **b** 3:7
c 35:15 **d** 49:21

15 Which ratio is *not* equivalent to 27:18?

a 3:2 **b** 9:6
c 54:36 **d** 18:27

16 How far would you ride in 2.5 hours at 11 km per hour?

a 18 km **b** 25 km
c 22 km **d** 27.5 km

17 If 38 litres of petrol is used to travel 361 km, how many kilometres are travelled per litre?

a 10 km **b** 9.5 km
c 3.61 km **d** 3.8 km

18 If 13.5 L of water is used per minute, how many litres are used in one hour?

a 135 L **b** 810 L
c 81 L **d** 1350 L

Assessment — Number and Application

Directed numbers and indices

1 Insert the symbol < or > to make a true statement.

a $+12 \ \square \ -5$ **b** $-8 \ \square \ -11$ **c** $-4 \ \square \ 0$ **d** $15 \ \square \ +13$
e $-34 \ \square \ -32$ **f** $+60 \ \square \ 67$ **g** $46 \ \square \ -6$ **h** $-81 \ \square \ 80$

2 Solve these additions and subtractions.

a $8 - -3$ **b** $+7 + -2$ **c** $-10 + +8$ **d** $-12 - +7$
e $+15 - +10$ **f** $-11 - -4$ **g** $23 - +14$ **h** $-30 + -10$
i $+7 - -6 + +2$ **j** $12 + -3 - +4$ **k** $-16 - +7 + -11$ **l** $21 - -6 - -8$
m $-18 + +15 - -9$ **n** $-14 + -0 - -6$ **o** $35 - -8 - +12$ **p** $-21 + +14 + -5$

3 Solve these multiplications and divisions.

a -6×-5 **b** 8×-7 **c** $-12 \times +6$ **d** $+8 \times 9$
e $4 \times -5 \times -2$ **f** $-3 \times -5 \times 10$ **g** $6 \times -4 \times -3$ **h** $8 \times -6 \times +2$
i $-24 \div -6$ **j** $60 \div -10$ **k** $-45 \div +5$ **l** $36 \div +9$
m $+55 \div 11$ **n** $-56 \div -8$ **o** $64 \div -8$ **p** $+32 \div 4$

4 Find the prime factors of each number and write them in index form. For example: $36 = 3 \times 3 \times 2 \times 2 = 3^2 \times 2^2$

a 20 **b** 48 **c** 100 **d** 75 **e** 80 **f** 500

5 Calculate these.

a $5^2 + 7^2 + 2^3$ **b** $12^2 - 6^2$ **c** $10^3 - 9^2$ **d** $4^3 + 8^2 - 3^3$
e $4^2 \times 2^3$ **f** $9^2 \div 3^2$ **g** $11^2 \times 2^3$ **h** $10^3 \div 5^2$
i $2^4 + 3^4$ **j** $10^3 - 5^4$ **k** $2^6 + 3^4 + 10^1$ **l** $8^4 - 11^0$
m $4^4 + 9^1 + 6^3$ **n** $7^5 - 3^6$ **o** $12^0 + 15^1 + 2^7$ **p** $9^4 - 3^5$

6 Calculate these.

a $\sqrt{64}$ **b** $\sqrt[3]{1000}$ **c** $\sqrt{81}$ **d** $\sqrt[3]{512}$ **e** $\sqrt{441}$ **f** $\sqrt[3]{27\,000}$
g $\sqrt{361}$ **h** $\sqrt[3]{15\,625}$ **i** $\sqrt{1600}$ **j** $\sqrt[3]{2744}$ **k** $\sqrt{484}$ **l** $\sqrt[3]{9261}$

7 Solve these multiplications and divisions.

a $7^0 \times 6^1$ **b** $12^1 \times 8^0$ **c** $10^0 \times 8^1 \times 11^0$ **d** $4^2 \times 5^0 \times 5^1$
e $8^2 \div 8^1$ **f** $2^4 \div 4^1$ **g** $28^1 \div 7^1$ **h** $42^1 \div 8^0$
i $4^3 \times 2^0 \times 2^2$ **j** $144^1 \div 12^2$ **k** $6^2 \times 2^3 \times 20^0$ **l** $6^3 \div 8^1$

Strand	Space and Shape

8.2.5 Investigate the area of circles

Calculate the area of circles

Help Box

The area of the circle shown here can be calculated using the formula $A = \pi r^2$, where A is the area, r is the radius and π represents 3.14 or $\frac{22}{7}$.

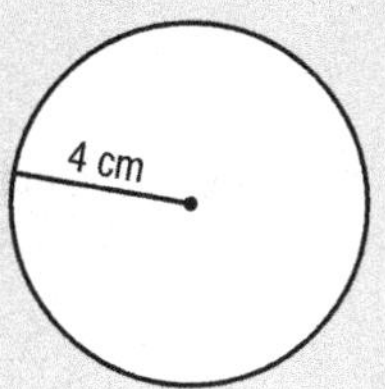

The area is 50.24 cm².

The radius is 4 cm so $A = 3.14 \times 4^2$ OR $A = \frac{22}{7} \times 4^2$

$= 3.14 \times 16$ $\quad$ $= \frac{22}{7} \times 16$

$= 50.24$ cm². $\quad$ $= \frac{352}{7}$

$= 50\frac{2}{7}$ cm².

If the diameter is given instead of the radius, halve the diameter before applying the formula, as the radius is half the length of the diameter.

Use π as either 3.14 or $\frac{22}{7}$ to calculate the area of these circles.

1

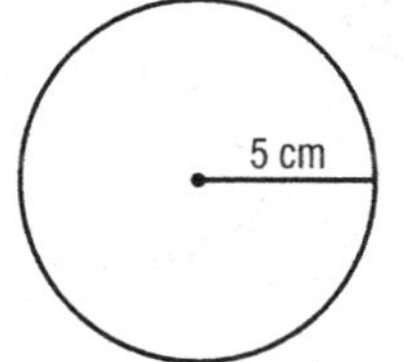

2

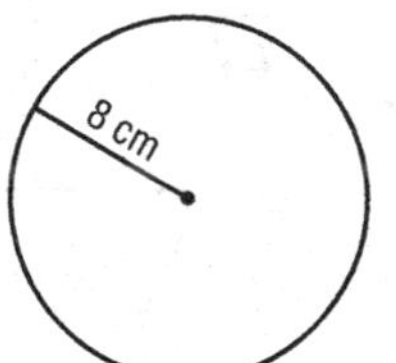

3

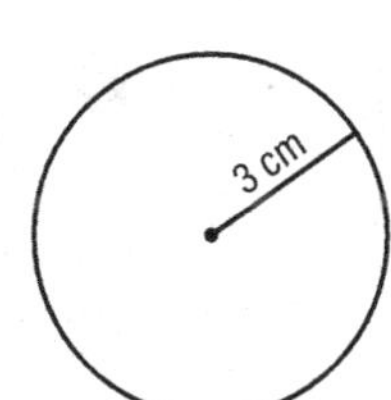

4

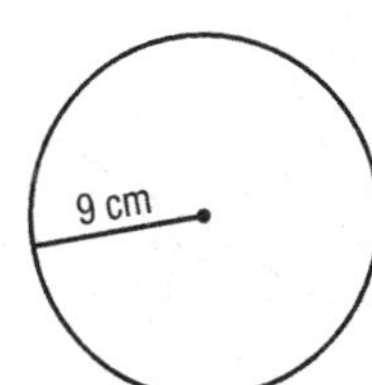

5

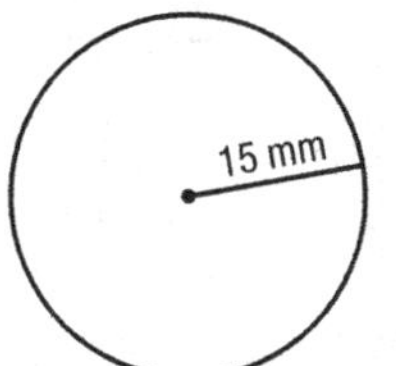

6

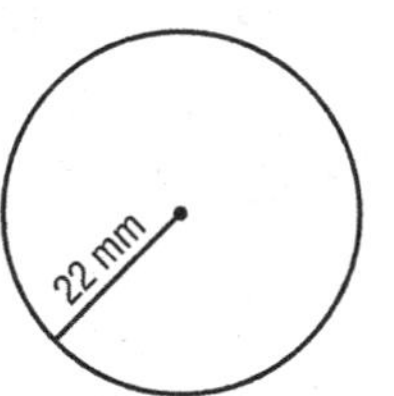

7

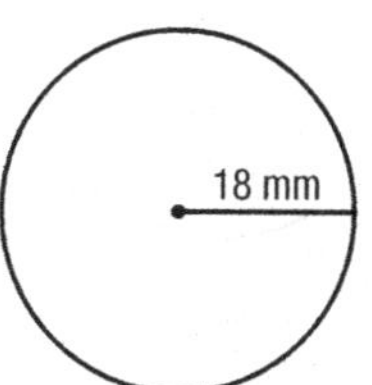

8

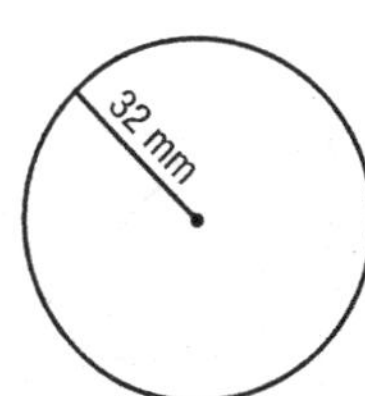

9

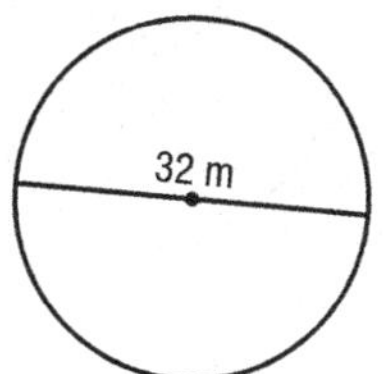

10

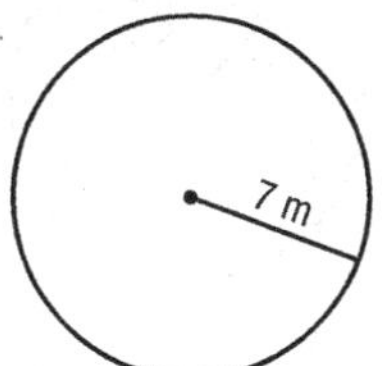

11

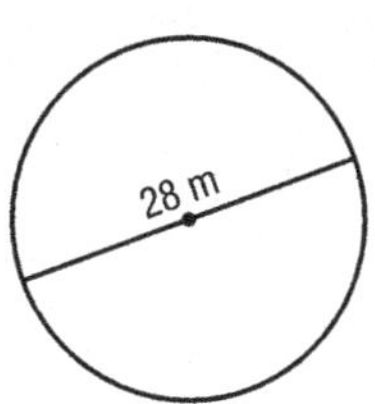

12

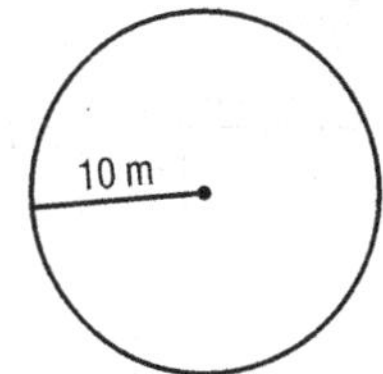

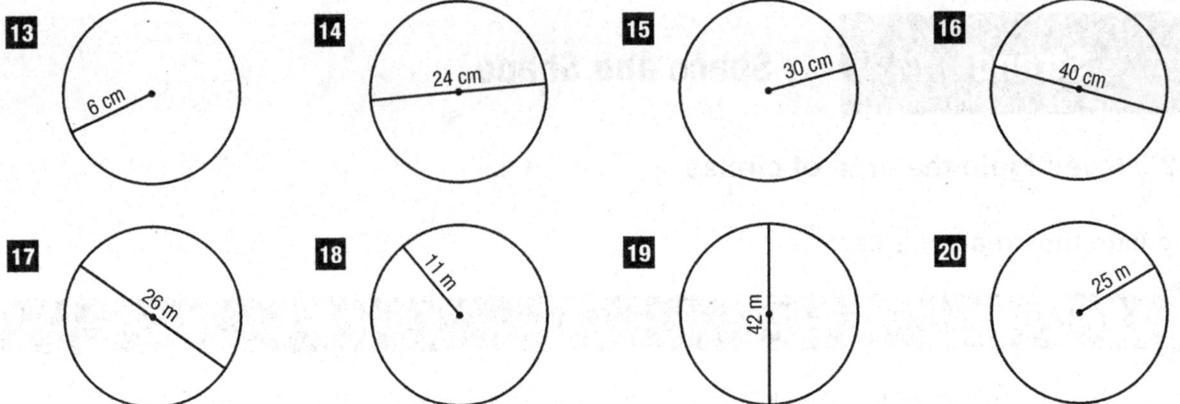

Calculate the area of part circles

Determine the part of each circle that is drawn ($\frac{1}{4}, \frac{1}{2}, \frac{3}{4}$) and then use what you know about the area of circles to calculate the area of each part, correct to two decimal places.

1. 4 cm
2. 7 cm
3. 12 cm
4. 6 cm
5. 10 cm
6. 8 cm
7. 3 m
8. 4 m
9. 9 m
10. 2 m
11. 24 m
12. 15 m
13. 25 mm
14. 45 mm
15. 21 mm
16. 36 mm
17. 30 mm
18. 80 mm

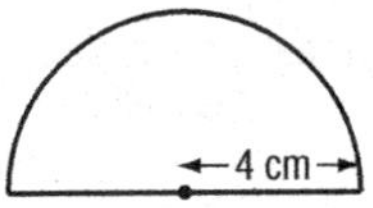

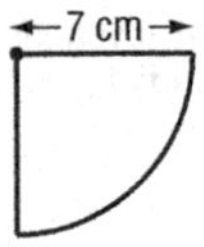

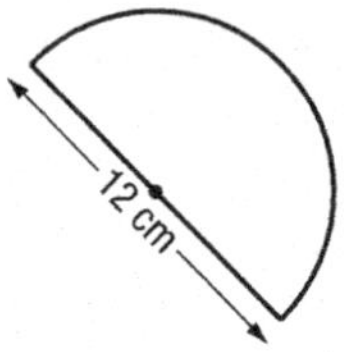

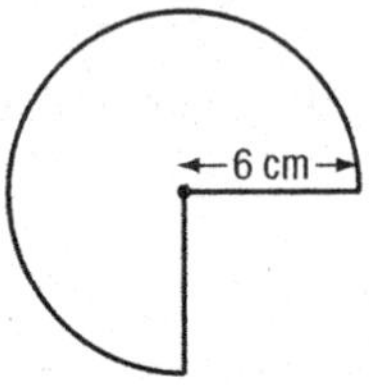

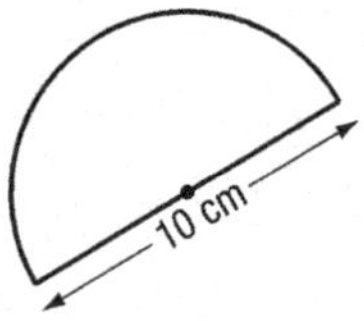

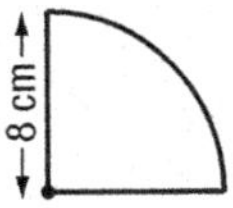

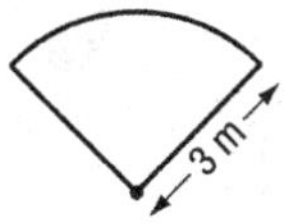

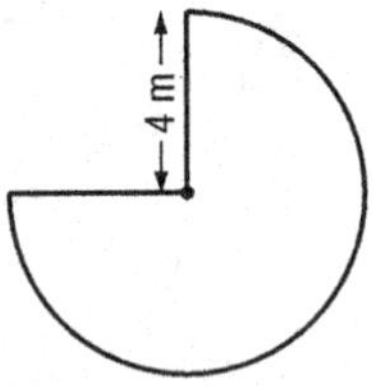

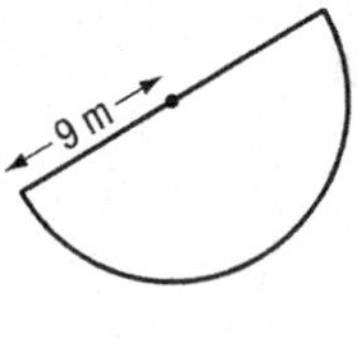

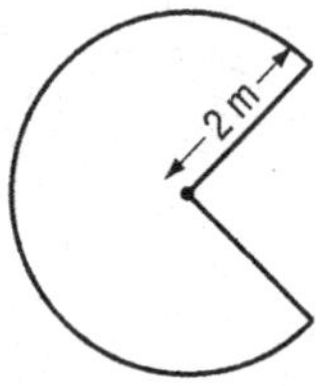

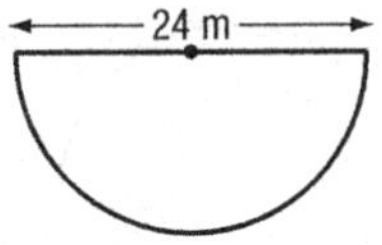

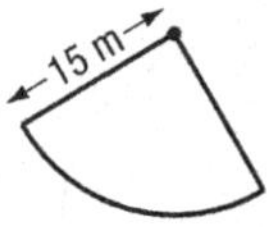

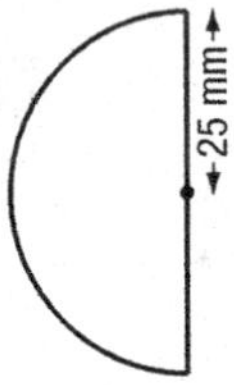

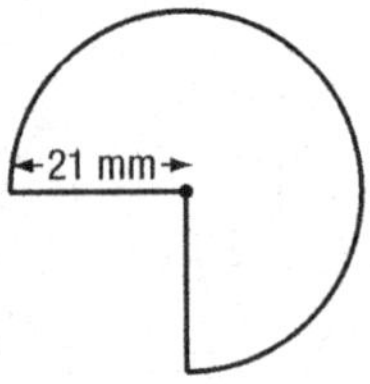

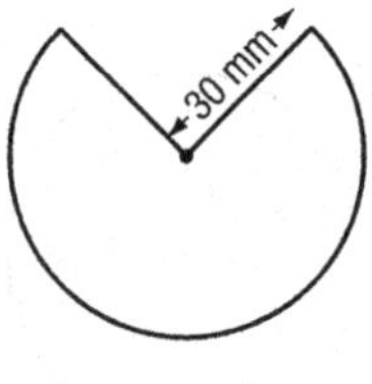

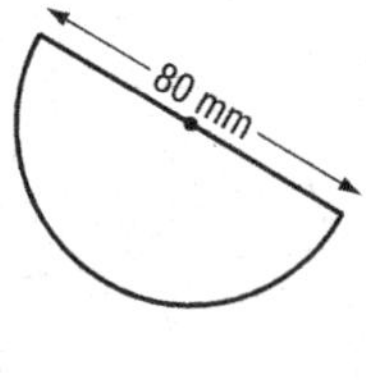

Calculate the area of shapes that include parts of circles

Calculate the area of each shape by finding the area of the rectangle and the area of the part circle and adding them together. Give your answers correct to two decimal places.

1

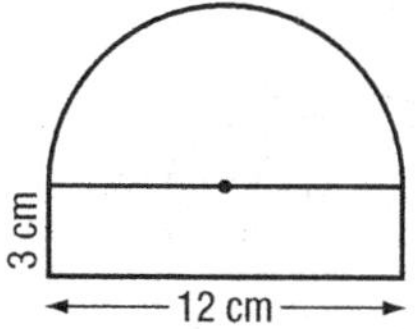

2

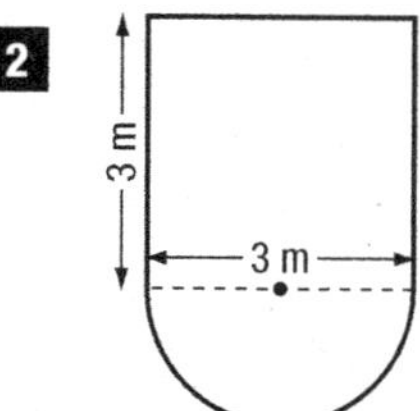

3

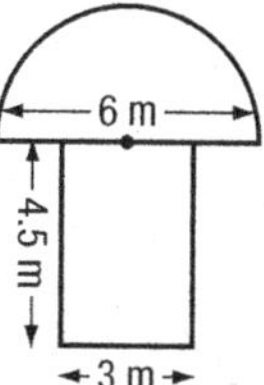

4

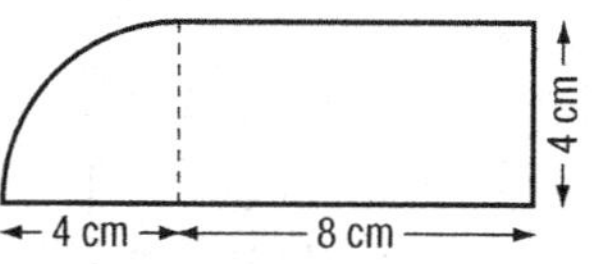

5

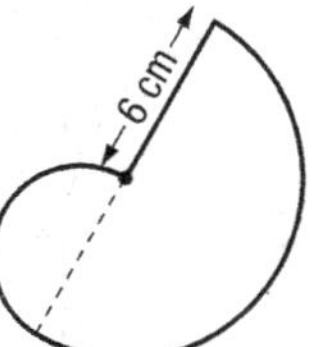

6

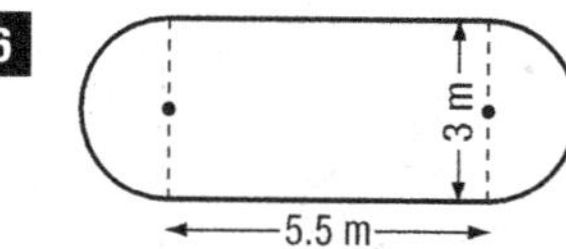

7

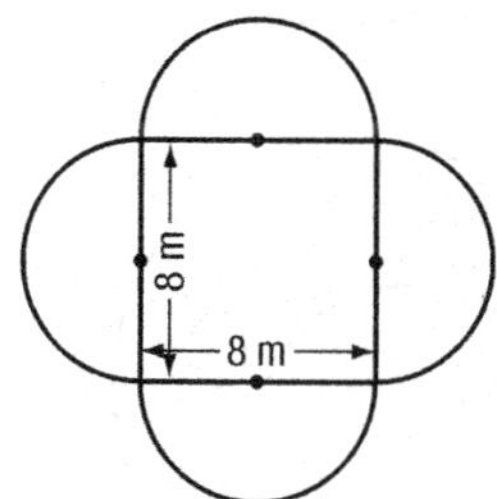

8

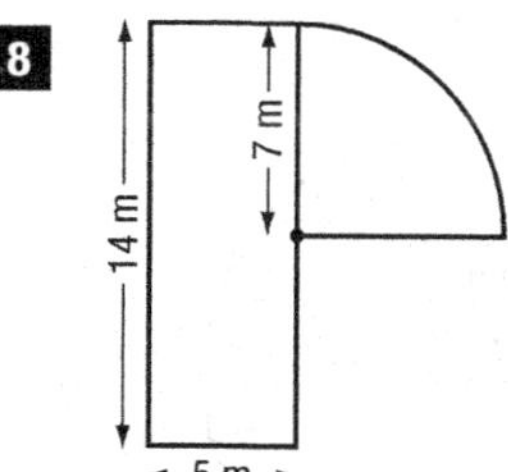

9

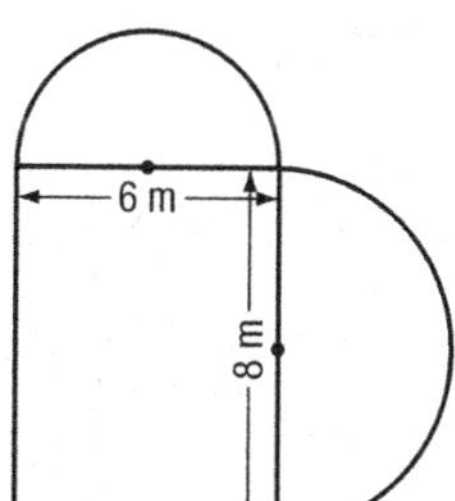

10

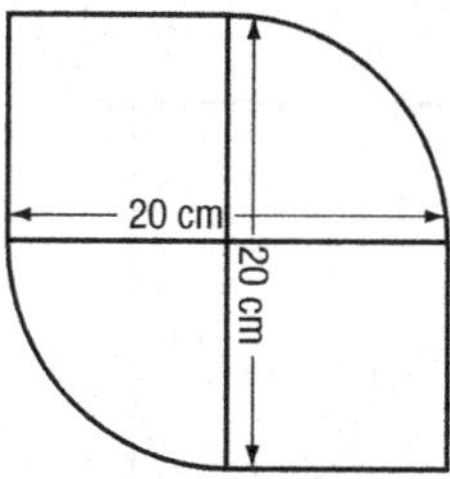

11

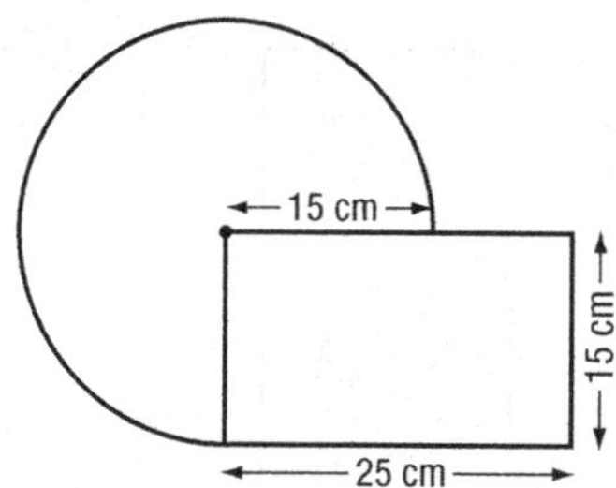

12

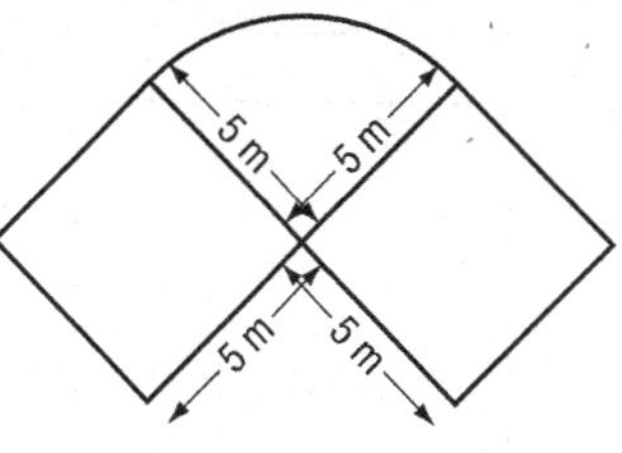

13

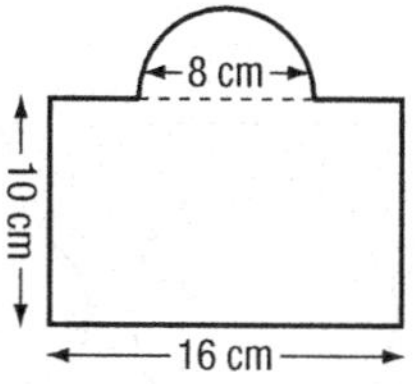

14

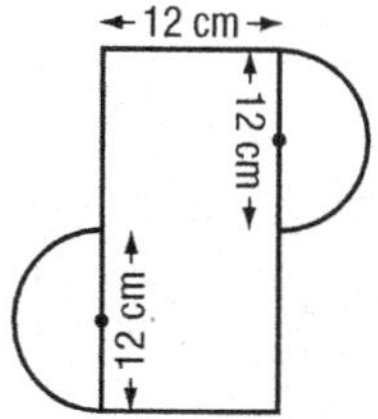

15

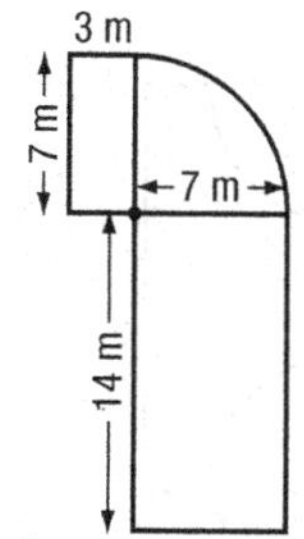

Calculate areas within and around circles

Calculate the shaded area of these diagrams by subtracting the area of the inner shape/s from the total area of the outside shape. Give your answers correct to two decimal places.

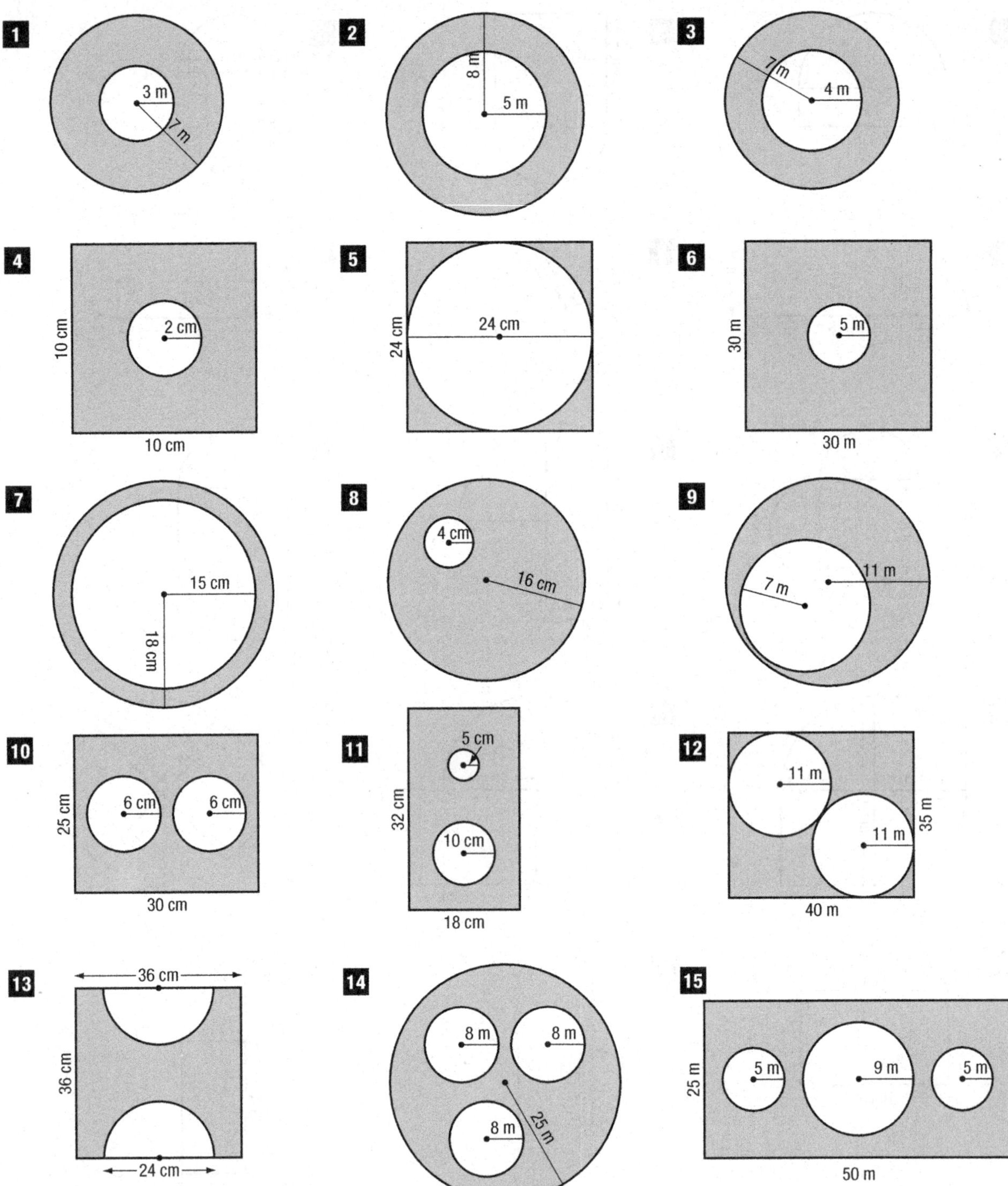

8.2.6 Investigate volumes of cylinders, cones and pyramids and apply some volume rules

Revise volume of rectangular prisms

Remember

The formula for finding the volume of a rectangular prism is: Length × Width × Height

Use the given measurements to calculate the volume of each rectangular prism.

1 4 cm, 10 cm, 4 cm

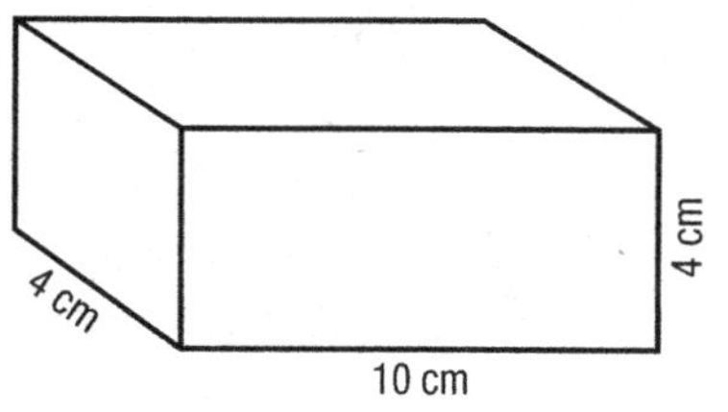

2 3 cm, 3 cm, 8.5 cm

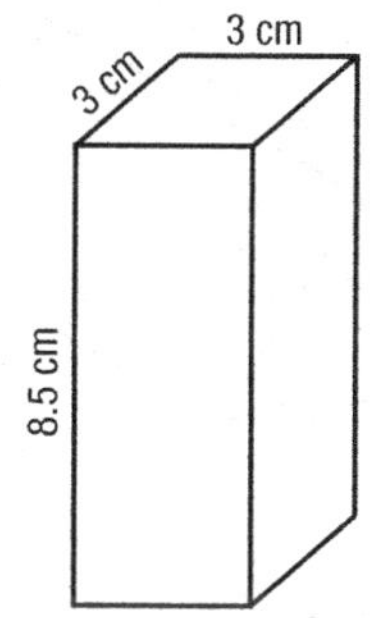

3 7 cm, 9.5 cm, 6 cm

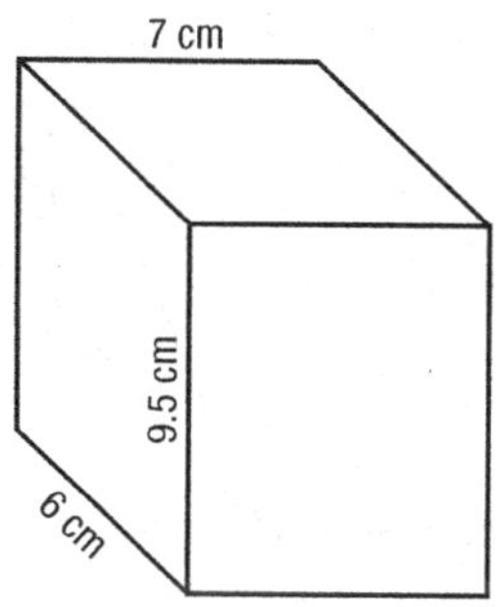

4 2.5 m, 3.5 m, 5 m

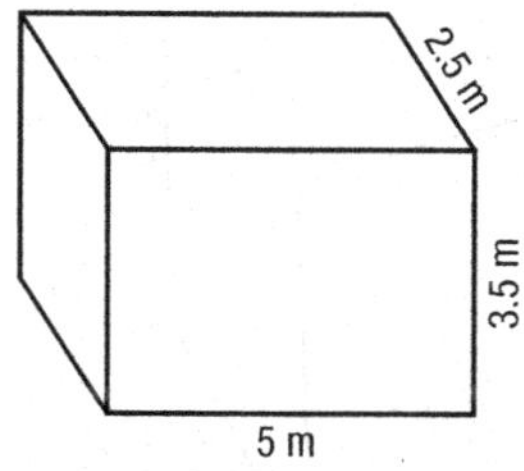

5 4 m, 3.5 m, 6 m

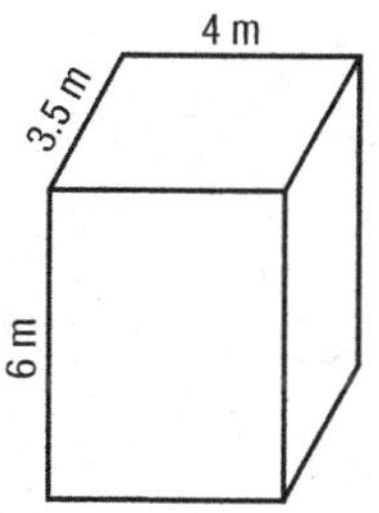

6 5 m, 5 m, 12.5 m

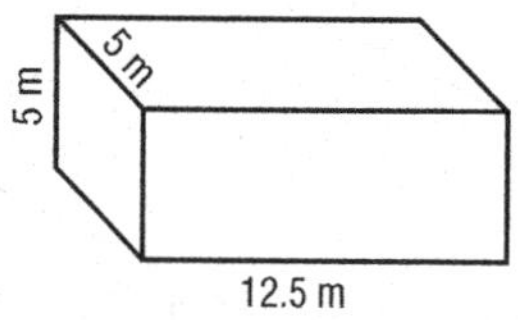

7 2.5 cm, 2.5 cm, 15 cm

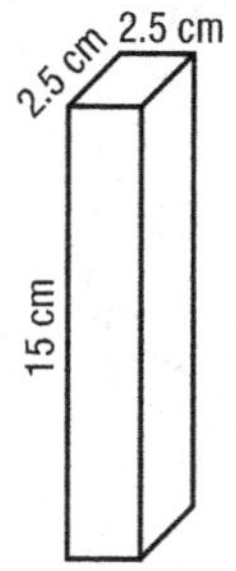

8 30 cm, 4.5 cm, 25 cm

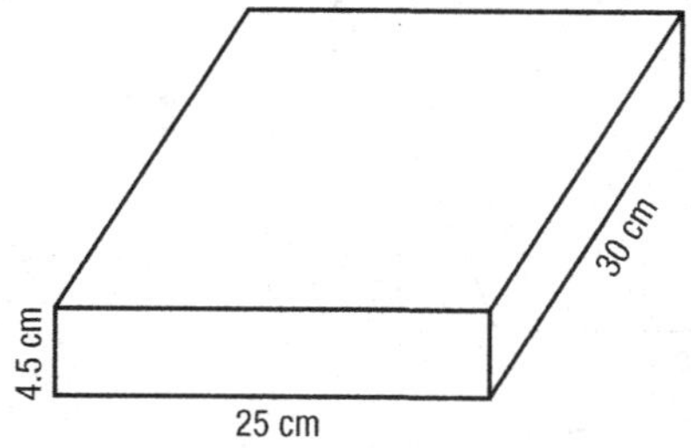

9 12 cm, 18 cm, 6 cm

10 6 m, 7.5 m, 7.5 m

11 9 m, 5.5 m, 6 m

12 11.5 m, 5 m, 5.5 m

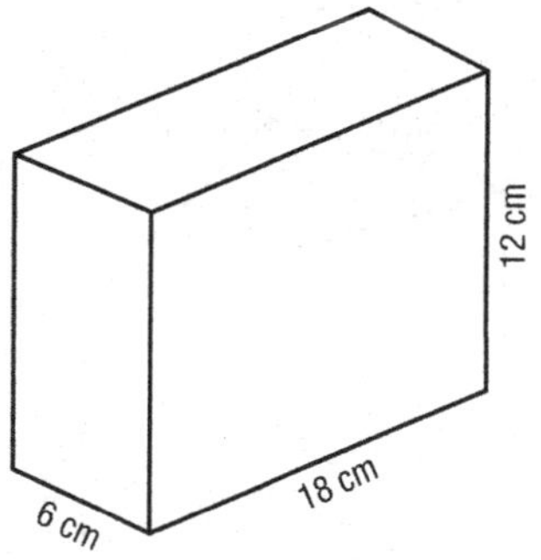

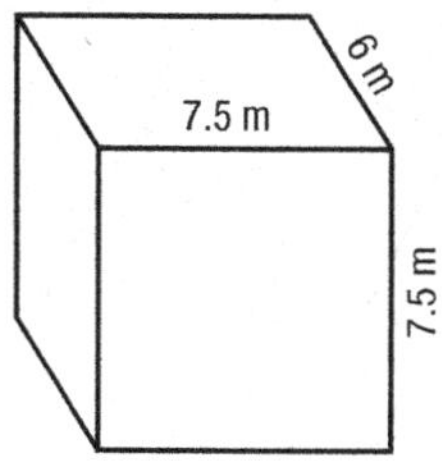

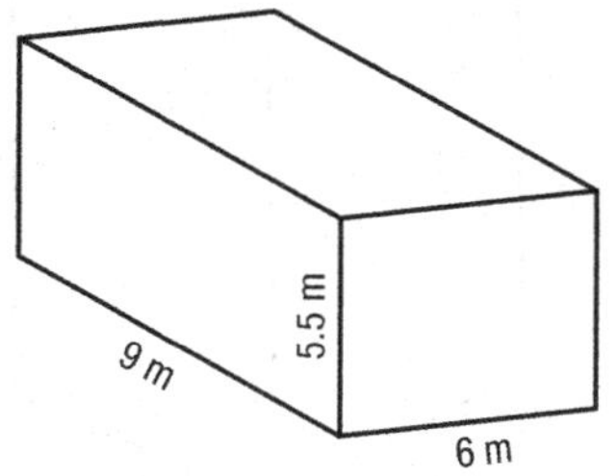

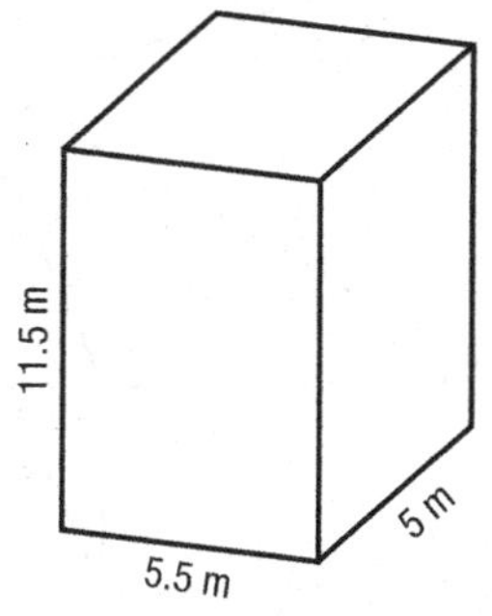

Revise volume of triangular prisms

Remember

To find the volume of a triangular prism:

1. calculate the area of the triangular base $\frac{(\text{Length} \times \text{Width})}{2}$
2. multiply the area of the base by the height of the prism (Area × Height)

Use the given measurements to calculate the volume of each triangular prism.

1 17 cm, 9 cm, 15.5 cm

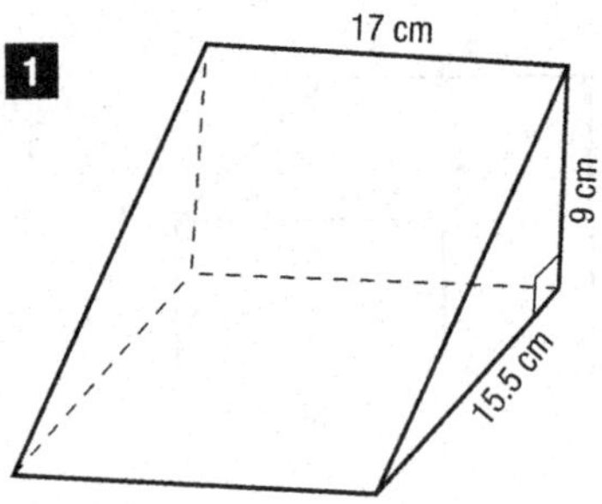

2 27.5 cm, 20 cm, 18 cm

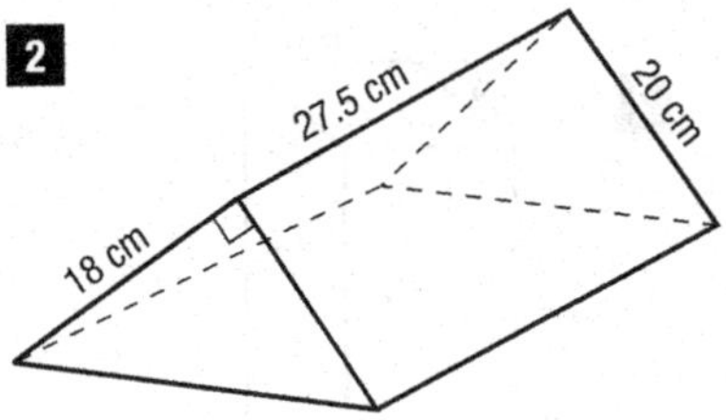

3 35 cm, 32 cm, 16 cm

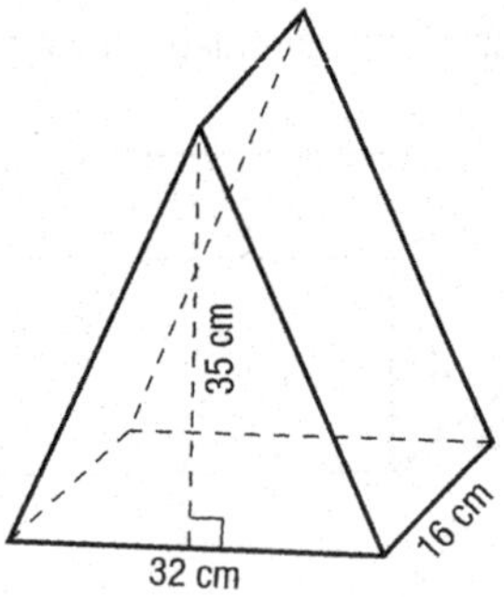

4 13 m, 7.5 m, 28 m

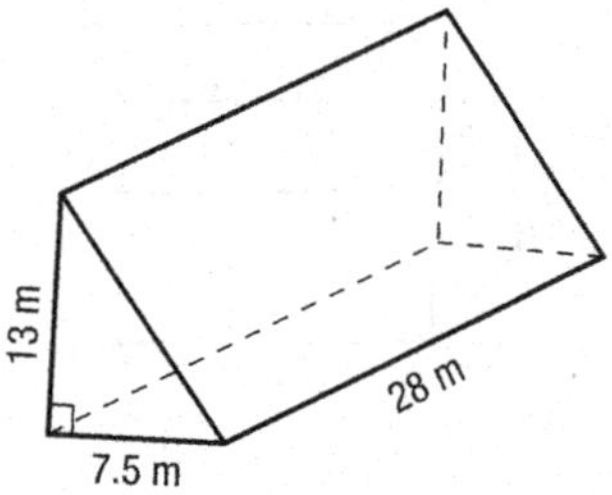

5 21 m, 30 m, 15 m

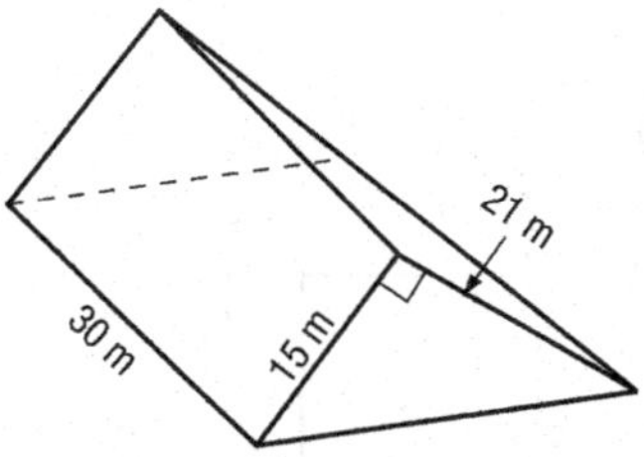

6 15 m, 12 m, 8.5 m

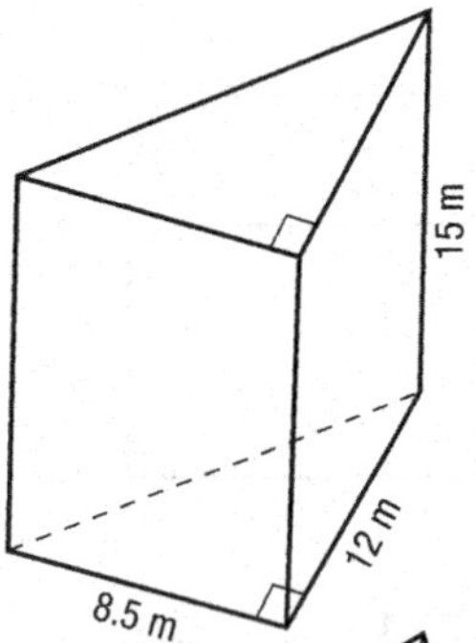

7 20 cm, 22 cm, 26 cm

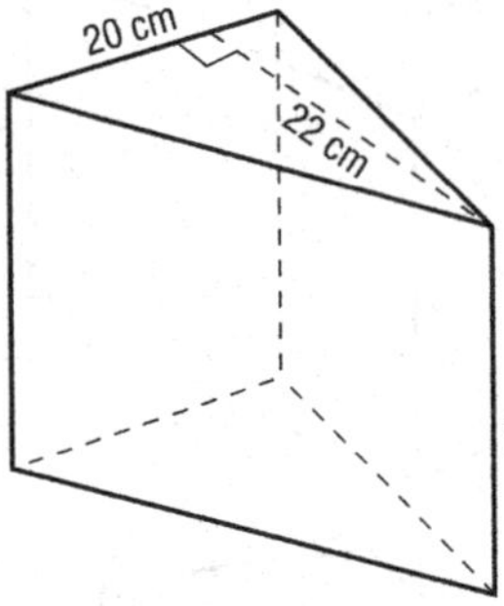

8 9 cm, 12 cm, 16 cm

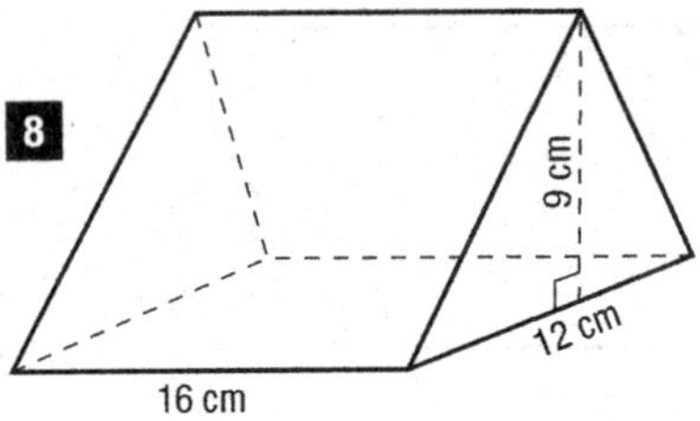

9 15 cm, 11 cm, 7 cm

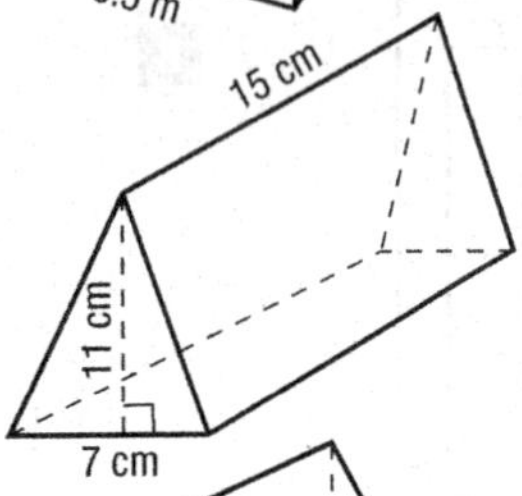

10 4 m, 9.5 m, 8.5 m

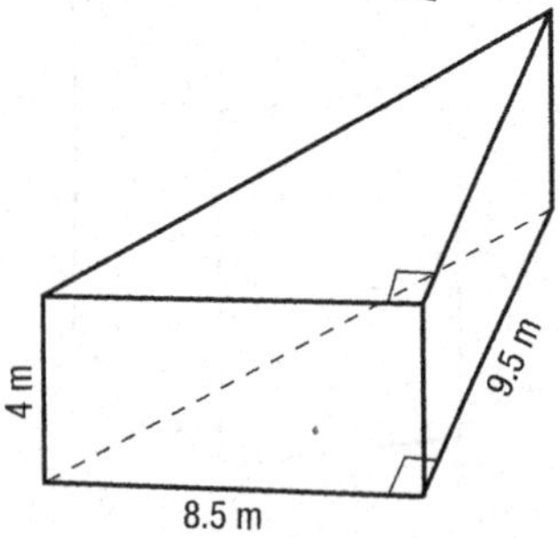

11 14.5 m, 7 m, 7 m

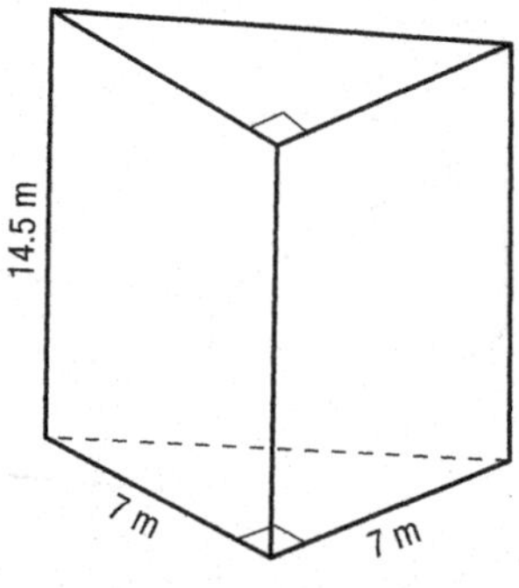

12 11 m, 6 m, 8 m

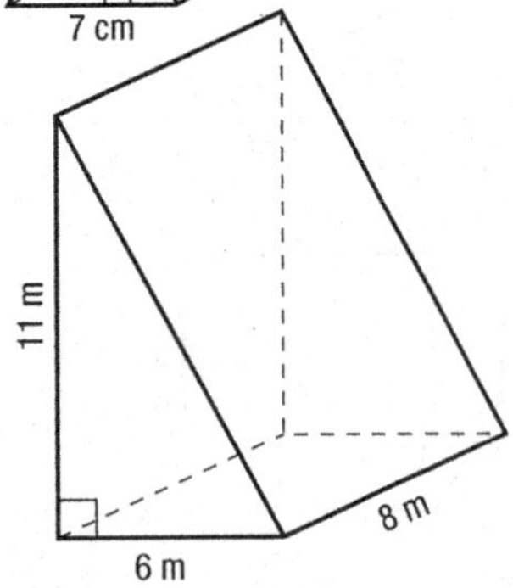

Calculate volume of cylinders

Help Box

To find the volume of a cylinder:

1. Calculate the area of the circular base ($A = \pi r^2$).
2. Multiply the area of the circular base by the height of the cylinder ($A \times h$).

The formula can be written as: V (volume) $= \pi r^2 h$ where r stands for radius, h stands for height and π represents 3.14.

Example:

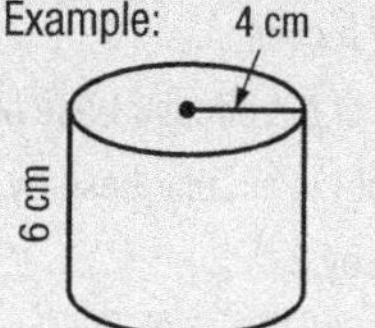

$3.14 \times 4^2 \times 6$
$= 3.14 \times 16 \times 6$
$= 301.44$ cm^3

The volume is 301.44 cm^3.

Use the given measurements to calculate the volume of each cylinder.

1

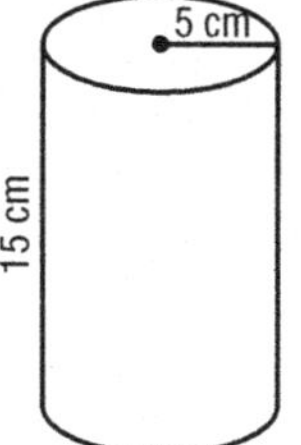

2

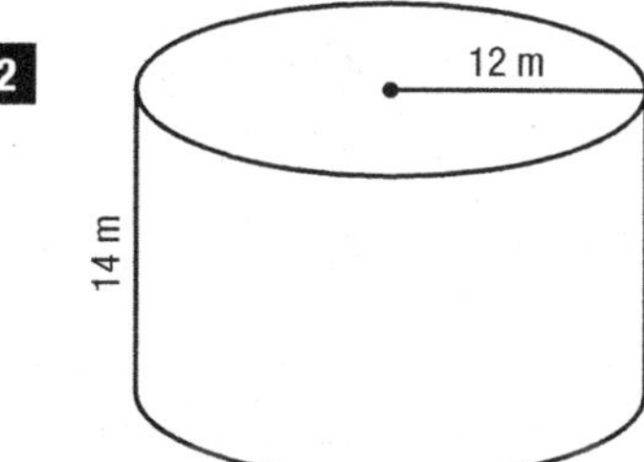

3

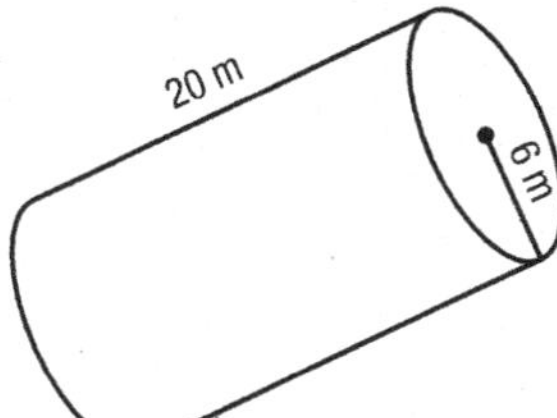

4

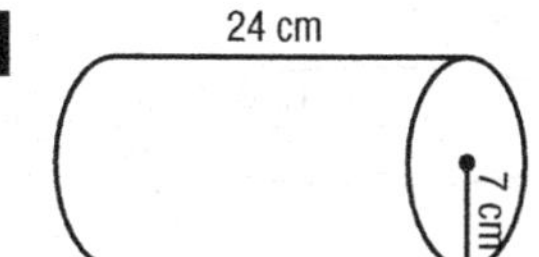

5

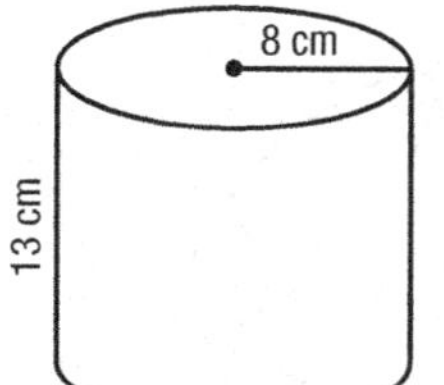

6

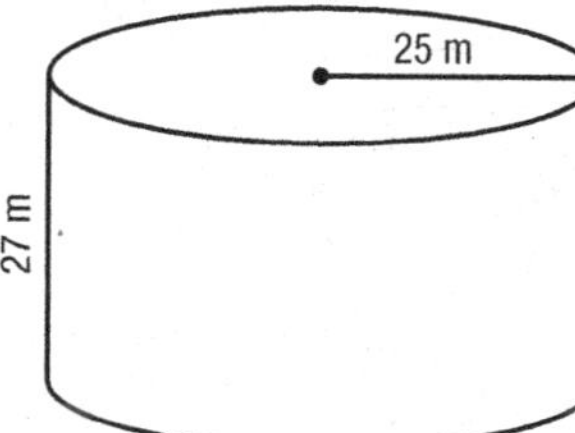

7

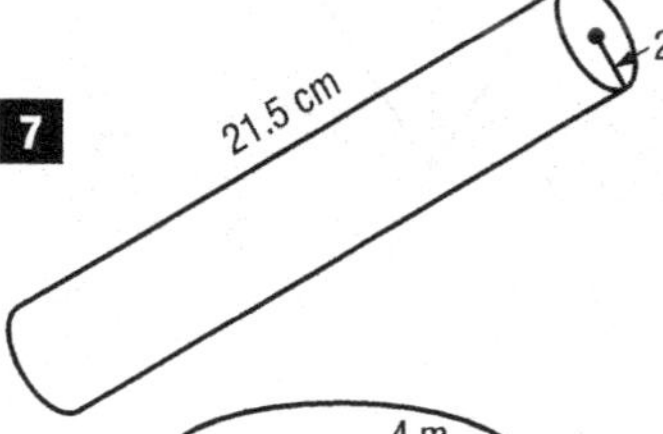

8

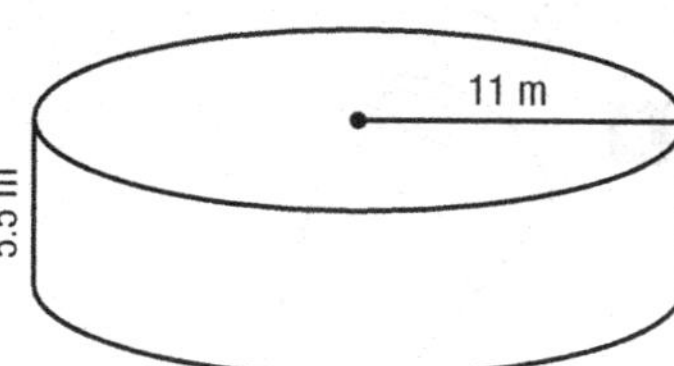

9

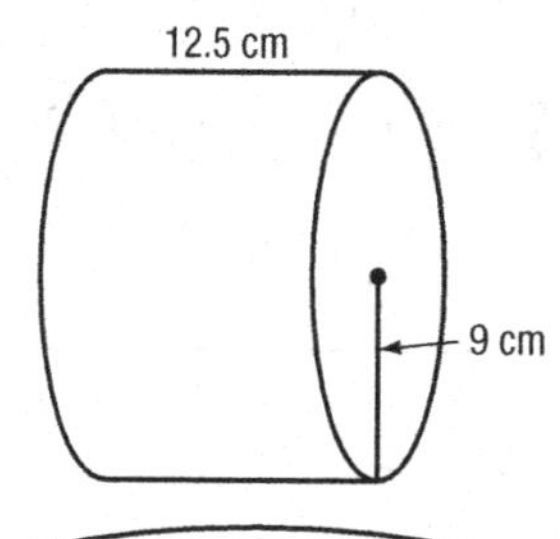

10

11

12

13

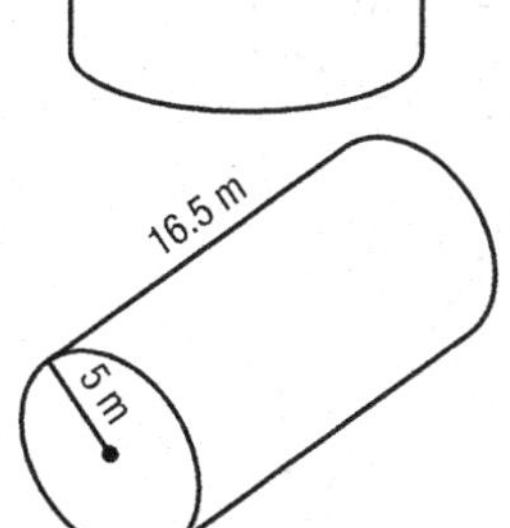

14

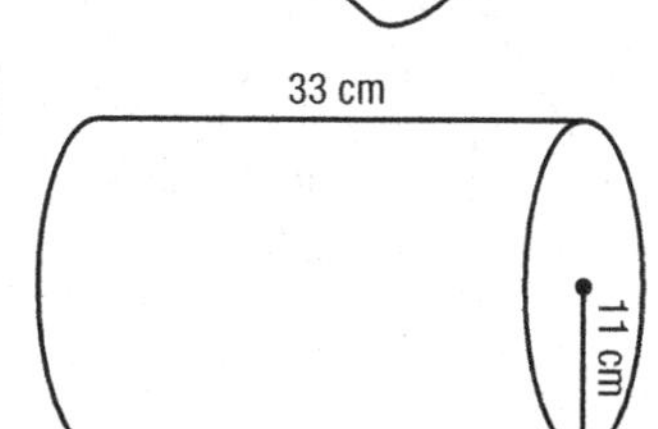

15

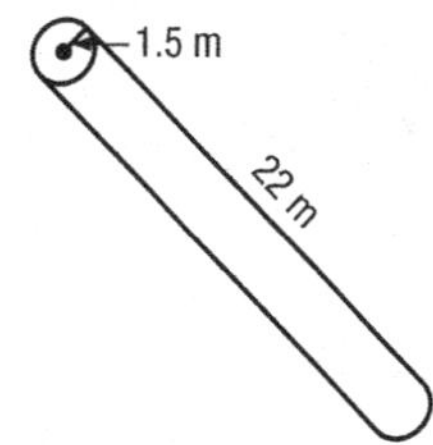

Calculate volume of cones

Help Box

To find the volume of a cone:

1. Calculate the area of the circular base ($A = \pi r^2$).
2. Multiply the area of the circular base by the height of the cone ($A \times h$).
3. Divide the product by 3.

The formula can be written as: V(volume) $= \frac{\pi r^2 h}{3}$ where r stands for radius, h stands for height and π represents 3.14.

Example:

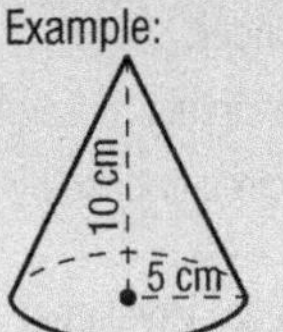

$$\frac{3.14 \times 5^2 \times 10}{3}$$
$$= \frac{3.14 \times 25 \times 10}{3}$$
$$= \frac{785}{3} = 261.67$$

The volume is 261.67 cm^3.

Use the given measurements to calculate the volume of each cone.

1

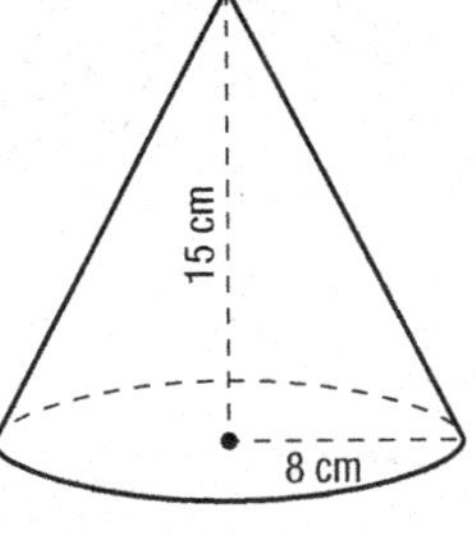

2

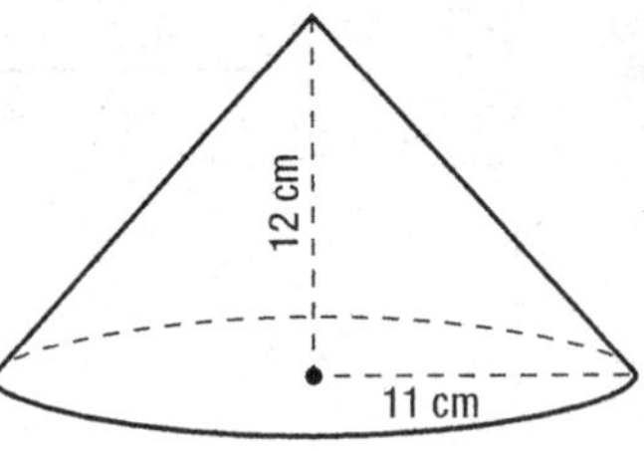

3

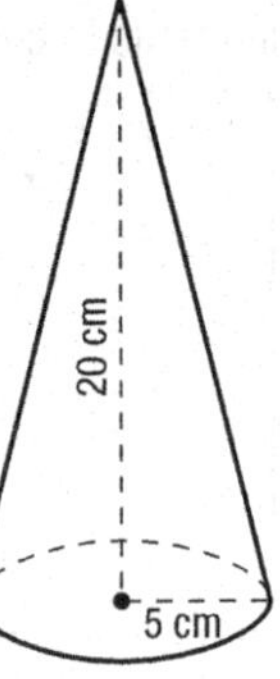

4

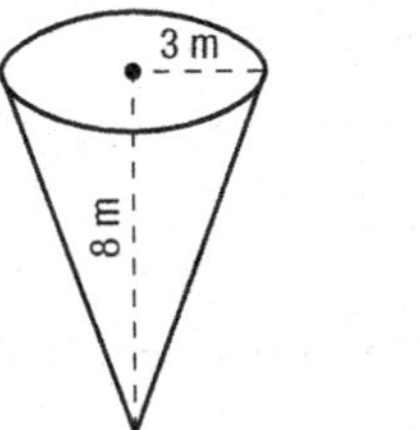

5

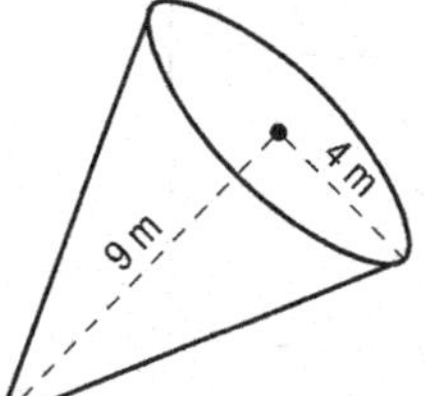

6

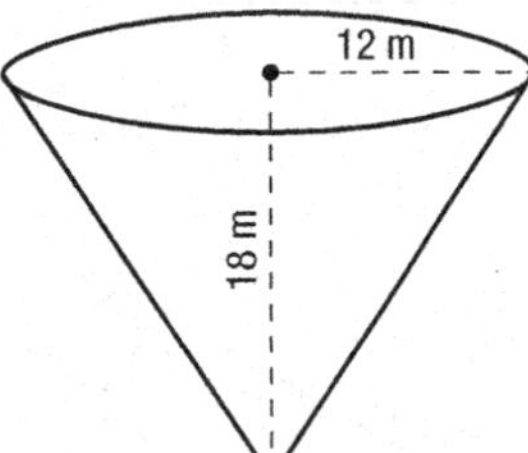

7

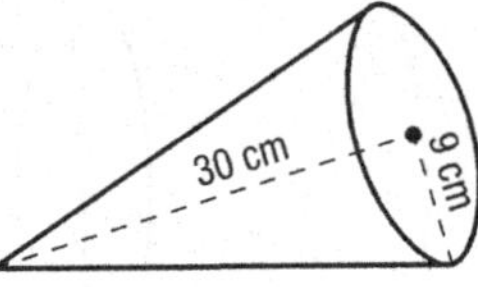

8

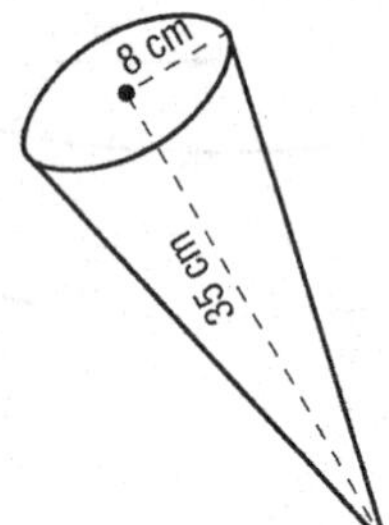

9

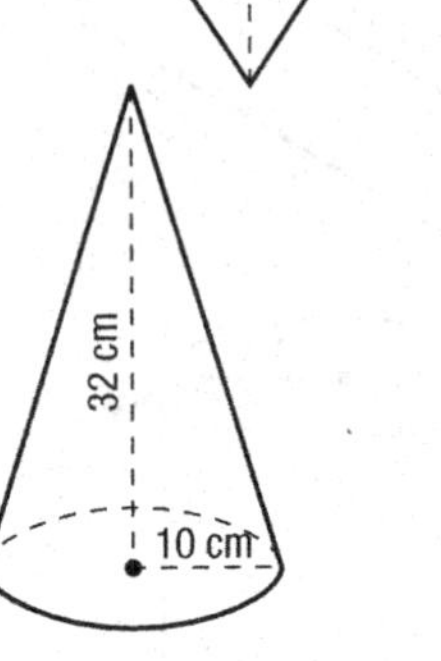

10

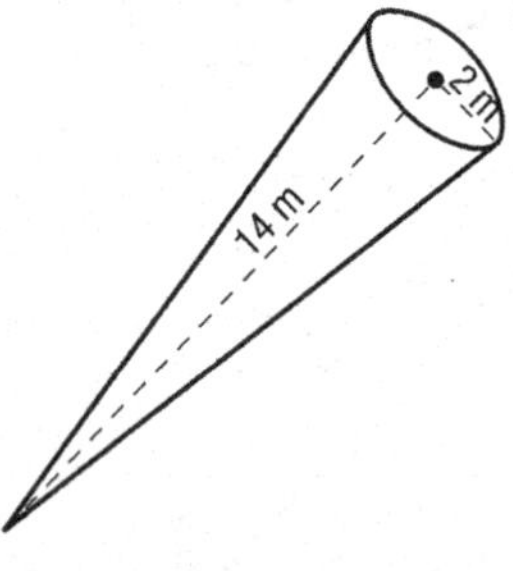

11

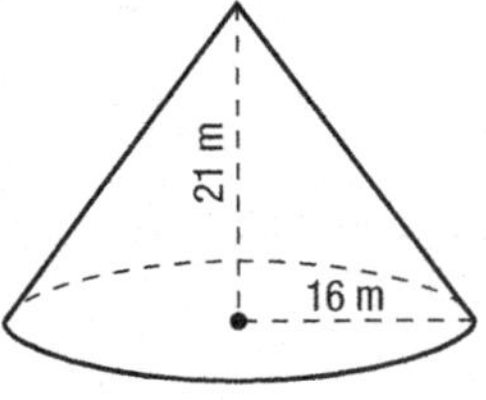

12

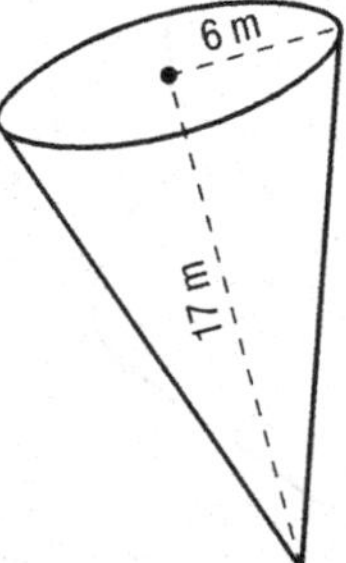

Calculate volume of pyramids

Help Box

To find the volume of a pyramid:

1. Calculate the area of the base.
2. Multiply the area of the base by the height of the pyramid ($A \times h$).
3. Divide the product by 3.

The formula can be written as: V (volume) $= \frac{Ah}{3}$ where A stands for area and h stands for height.

Example:

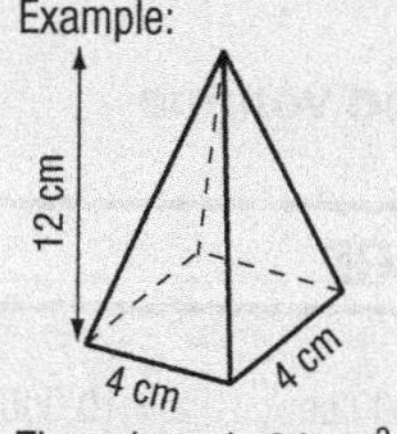

$\frac{4 \times 4 \times 12}{3}$

$= \frac{16\text{ cm}^2 \times 12\text{ cm}}{3}$

$= \frac{192}{3} = 64\text{ cm}^3$

The volume is 64 cm^3.

Look carefully at the base of each pyramid to see if it is square, rectangular or triangular and then use the given measurements to calculate the volume.

1

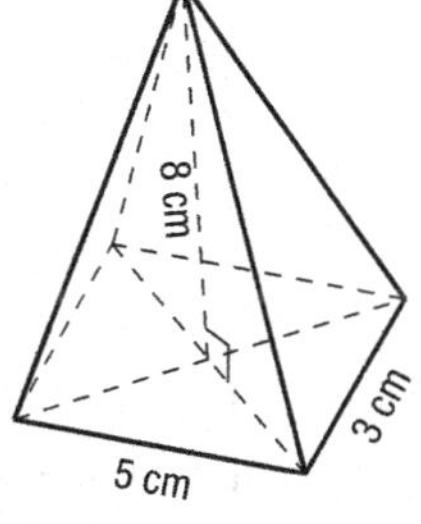

2

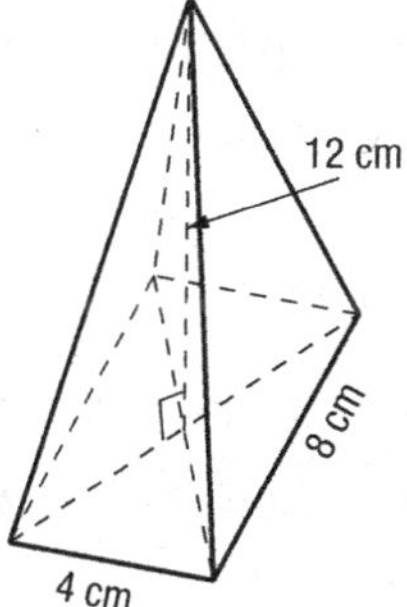

3

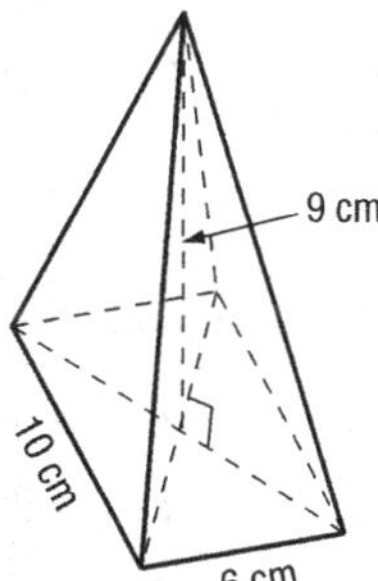

4

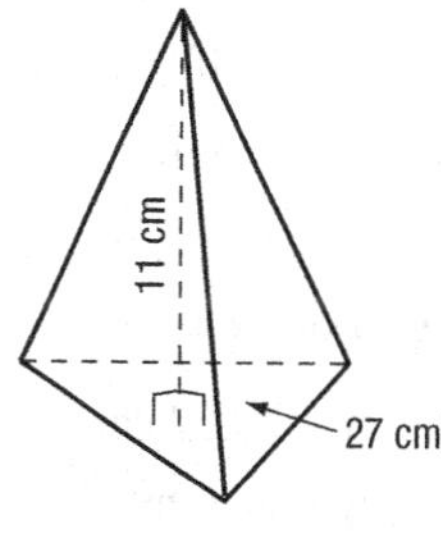

5

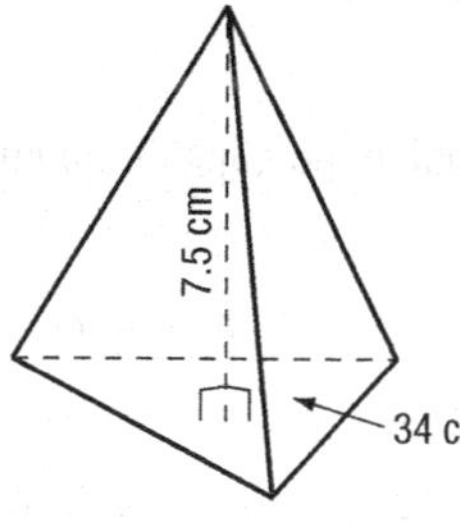

6

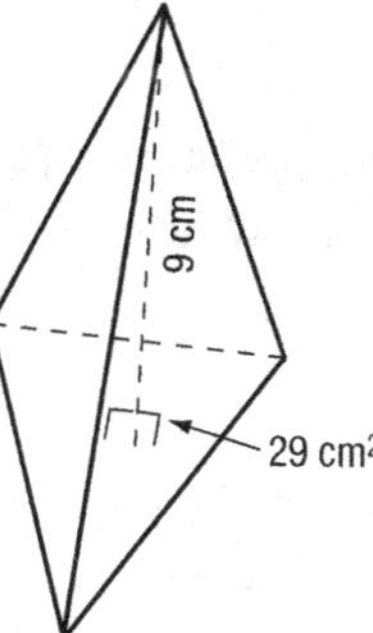

7

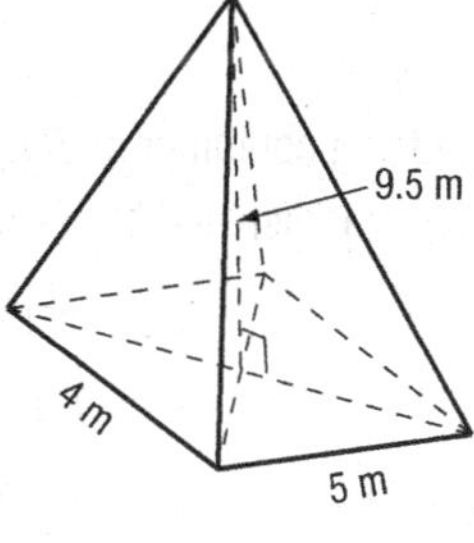

8

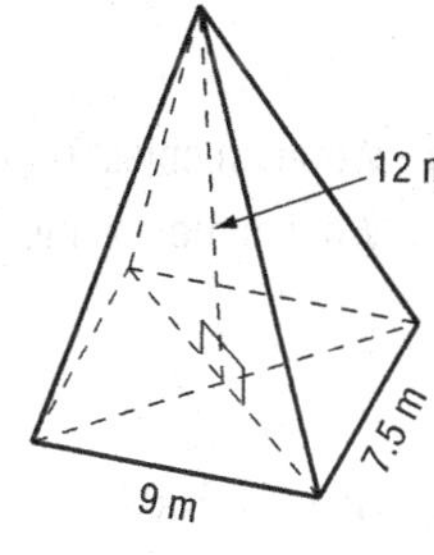

9

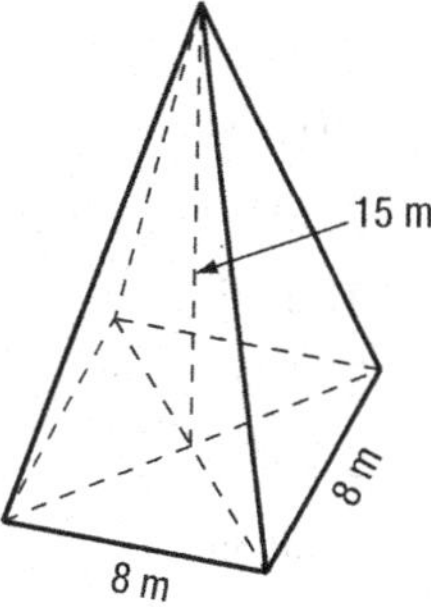

10

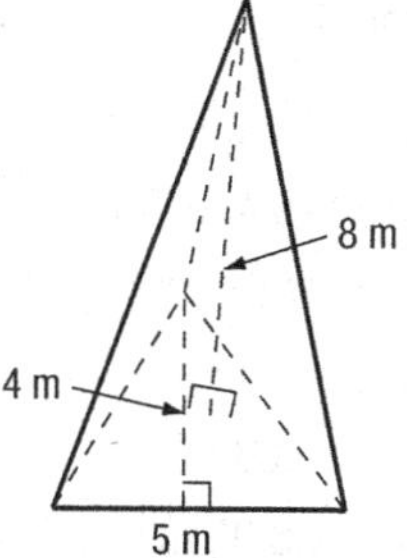

11

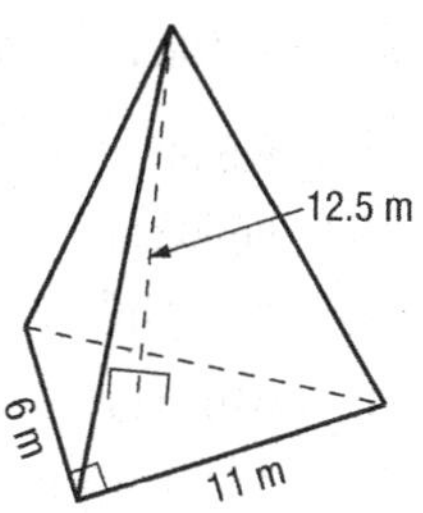

12

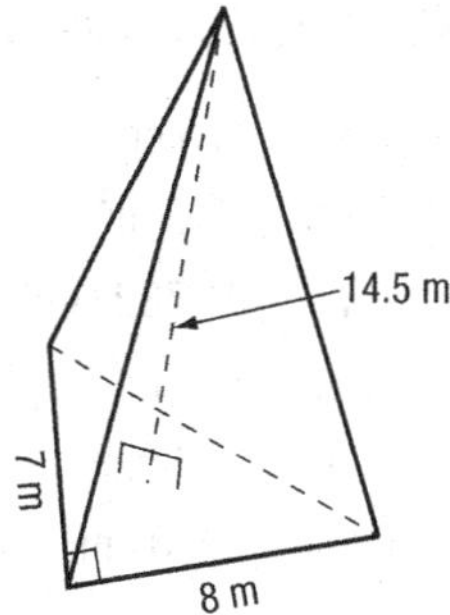

8.2.7 Apply capacity and volume measurements in problem solving

8.2.8 Convert between a variety of units: capacity and volume, and metric and imperial

Relate capacity and volume

Remember

1 mL = 1 cm^3; 1 L = 1000 cm^3; 10 L = 10 000 cm^3; 1000 L = 1 kL = 1 000 000 cm^3 = 1 m^3.

1 What would be the volume of containers with the following capacity?

a	30 mL	b	865 mL	c	7 L	d	3.75 L	e	2 kL	f	4.5 kL
g	3000 L	h	0.9 kL	i	187 mL	j	0.645 L	k	2.39 L	l	324 mL
m	8000 L	n	6.1 kL	o	40 000 L	p	1983 mL	q	518 mL	r	748 L
s	500 L	t	2075 L	u	15 kL	v	32.8 kL	w	325 L	x	3018 mL

2 What would be the capacity of containers with the following volume?

a	1327 cm^3	b	56 cm^3	c	4 m^3	d	11 m^3	e	8.5 m^3	f	1.75 m^3
g	5.4 m^3	h	0.7 m^3	i	450 cm^3	j	2716 cm^3	k	33 cm^3	l	6850 cm^3
m	12.9 m^3	n	6.57 m^3	o	543 m^3	p	4.18 m^3	q	917 cm^3	r	1056 cm^3
s	4387 cm^3	t	28 cm^3	u	115 m^3	v	1038 m^3	w	15.3 m^3	x	26.7 m^3

3 When an object is submerged in a container of water, the amount of water it displaces is equal to the object's volume. What is the volume of objects that displace the following amounts of water?

a	415 mL	b	10 kL	c	2500 L	d	78 mL	e	27 L	f	8250 L
g	16.25 kL	h	195 L	i	880 mL	j	11.25 L	k	5640 mL	l	36 500 L
m	17 750 L	n	45.2 kL	o	0.065 kL	p	934 mL	q	9.788 L	r	548 mL
s	1.23 kL	t	0.49 L	u	6.1 kL	v	14 300 L	w	400 mL	x	5.8 L

4 How much water would objects of the following volumes displace?

a	475 cm^3	b	8.5 m^3	c	0.54 m^3	d	19 cm^3
e	17.8 m^3	f	5690 cm^3	g	15 450 cm^3	h	36 m^3
i	31 732 cm^3	j	11.76 m^3	k	1093 cm^3	l	41.3 m^3
m	600 m^3	n	6460 cm^3	o	7 cm^3	p	23.1 m^3
q	2.96 m^3	r	13.09 m^3	s	0.03 m^3	t	0.4 m^3
u	7200 cm^3	v	17 835 cm^3	w	22.05 m^3	x	7.008 m^3

Solve word problems linking volume and capacity

1 A rectangular fish tank has a length of 50 cm, a width of 28 cm and a height of 35 cm. How many litres of water will it take to fill it to 5 cm below the top?

2 A rectangular tank has the following dimensions: 5 m × 6 m × 2.5 m. What is its capacity in kilolitres?

3 A large cylindrical tank has a base area of 7 m^2 and a height of 2.75 m. What is its capacity in kilolitres?

4 A swimming pool is 12 m wide, 25 m long and 1.6 m deep.

- **a** How many litres of water are needed to fill the pool?
- **b** If water costs 7.5t per litre, how much will it cost to fill the pool?

5 A small fish tank is in the shape of a rectangular prism with the following dimensions: 18 cm × 24 cm × 32.5 cm.

- **a** What is the volume of the fish tank?
- **b** What is the capacity of the tank in litres?

6 A circular pond has a diameter of 6 metres and a depth of 80 cm. Find the capacity of the pond to the nearest litre.

7 A cylindrical tin has a radius of 9 cm and a height of 45 cm. Find the capacity of the tin in litres, to one decimal place.

8 Draw and label the dimensions of two different rectangular tanks that will each hold 10 litres of water when full.

9 Calculate the height of a prism that has a base area of 12 m^2 and a capacity of 60 000 litres.

10 A small pool is three-quarters full of water. It measures 1.5 metres × 1.5 metres and is 50 cm deep. How much water is in the pool?

11 Which container of each pair has the greater capacity?

- **a** A circular pond with a capacity of 1150 litres or a circular pond with a radius of 1 metre and a height of 30 cm.
- **b** A rectangular tank with dimensions of 4 m × 5 m × 8 m or a rectangular tank with a capacity of 150 kilolitres.
- **c** A circular tin with a capacity of 750 mL or a rectangular tin with dimensions of 12 cm × 8 cm × 9 cm.

Convert between metric and imperial units

Help Box

Sometimes it is necessary to convert between imperial and metric measurements. The following conversions provide a suitable level of accuracy:

1 gallon = 4.5 L   1 foot = 0.3 m (30 cm)   1 inch = 25 mm   1 mile = 1.6 km

When converting imperial to metric use multiplication, for example: 23 gallons converted to litres is 23 multiplied by 4.5, so $23 \times 4.5 = 103.5$ litres; 45 feet converted to metres is 45 multiplied by 0.3, so $45 \times 0.3 = 13.5$ metres.

When converting metric to imperial use division, for example: 156 litres converted to gallons is 156 divided by 4.5, so $156 \div 4.5 = 34.67$ gallons; 25 metres converted to feet is 25 divided by 0.3, so $25 \div 0.3 = 83.33$ feet.

1 Convert these gallons to litres.

- **a** 12 gallons
- **b** 35 gallons
- **c** 89 gallons
- **d** 64 gallons
- **e** 128 gallons
- **f** 217 gallons
- **g** 151 gallons
- **h** 320 gallons

2 Convert these litres to gallons.

- **a** 180 litres
- **b** 565 litres
- **c** 310 litres
- **d** 275 litres
- **e** 450 litres
- **f** 748 litres
- **g** 516 litres
- **h** 602 litres

3 Which is the greater capacity?

- **a** 500 litres or 150 gallons
- **b** 85 gallons or 400 litres
- **c** 1000 litres or 225 gallons
- **d** 550 gallons or 2500 litres
- **e** 750 litres or 175 gallons
- **f** 330 gallons or 1500 litres
- **g** 920 litres or 200 gallons
- **h** 150 gallons or 645 litres
- **i** 325 litres or 75 gallons

4 Convert these imperial measurements to the given metric measurements.

- **a** 12 feet to metres
- **b** 9 inches to millimetres
- **c** 30 miles to kilometres
- **d** 24 inches to millimetres
- **e** 85 miles to kilometres
- **f** 50 feet to metres
- **g** 130 miles to kilometres
- **h** 45 feet to metres
- **i** 15 inches to millimetres

5 Convert these metric measurements to the given imperial measurements.

- **a** 80 kilometres to miles
- **b** 1000 millimetres to inches
- **c** 250 metres to feet
- **d** 50 metres to feet
- **e** 650 millimetres to inches
- **f** 260 kilometres to miles
- **g** 820 millimetres to inches
- **h** 145 kilometres to miles
- **i** 190 metres to feet

6 Which length in each pair is longer?

- **a** 50 feet or 16 metres
- **b** 30 inches or 600 mm
- **c** 20 km or 12 miles
- **d** 460 mm or 20 inches
- **e** 90 miles or 140 km
- **f** 35 metres or 110 feet
- **g** 300 km or 180 miles
- **h** 85 feet or 28 metres
- **i** 55 inches or 1250 mm

8.2.9 Make physical models of circles and investigate their properties

Calculate circumference of circles

Remember

To calculate circumference use either $C = \pi D$ or $C = 2\pi r$, where C is circumference, D is diameter, r is radius and π represents 3.14.

1 Calculate the circumference of the following circles by measuring either the radius or the diameter.

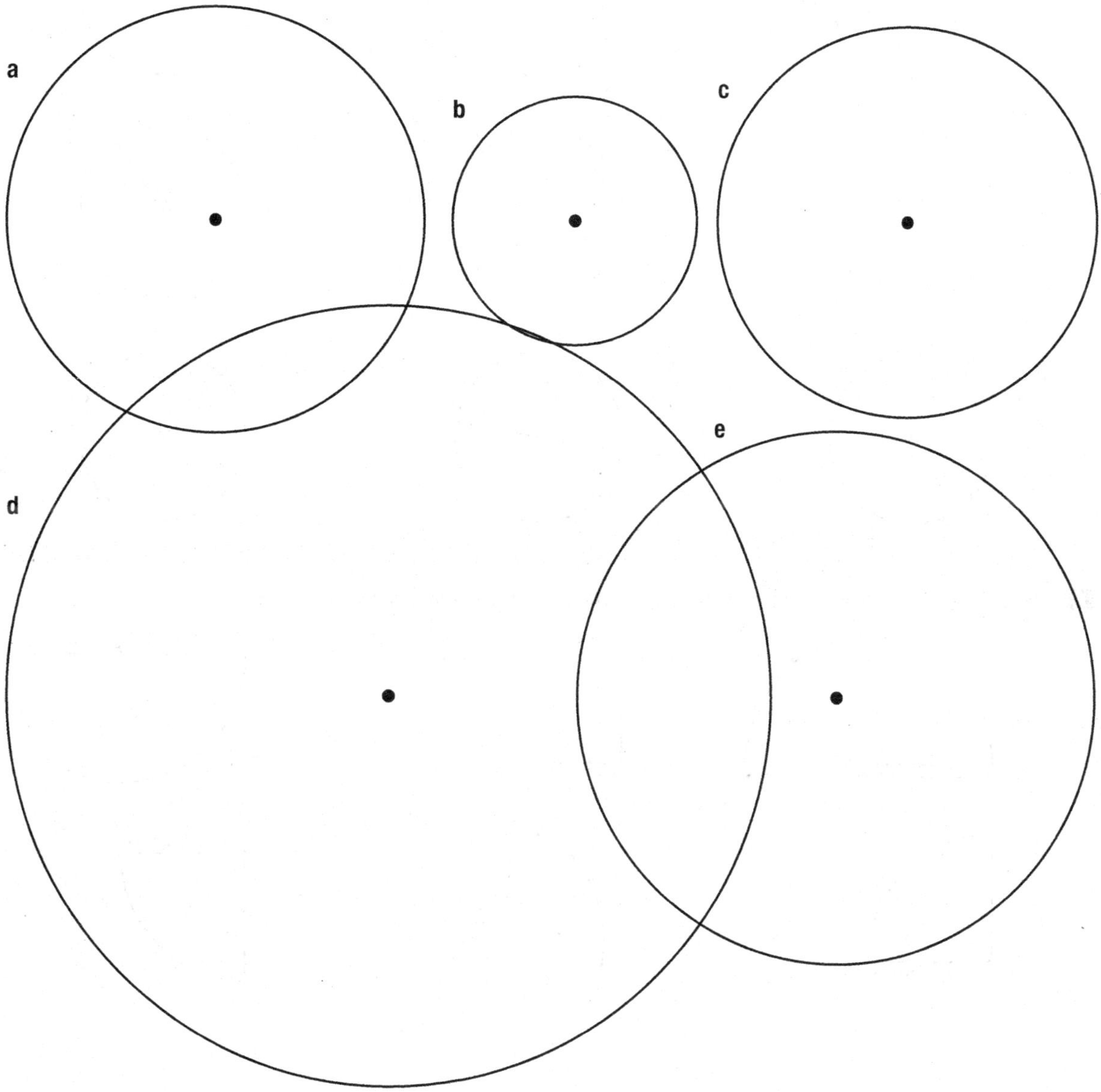

a

b

c

d

e

2 Calculate the circumference of each circle correct to two decimal places.

a

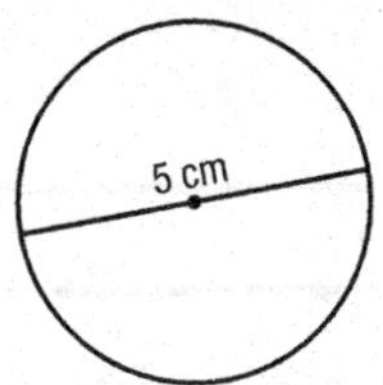

b

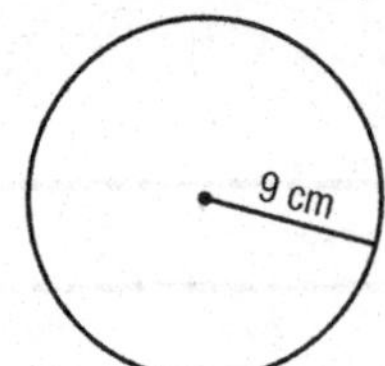

c

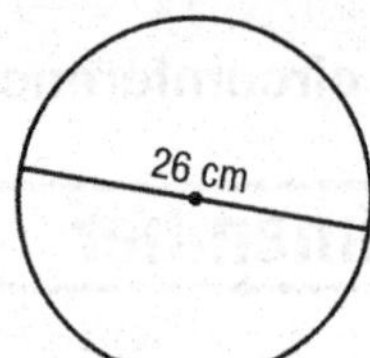

d

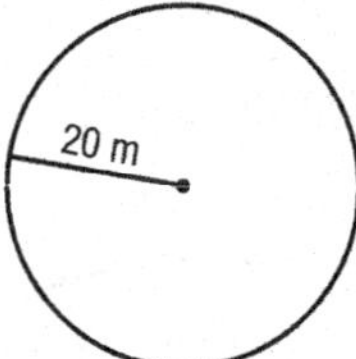

e

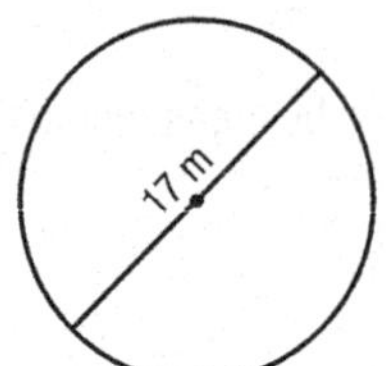

f

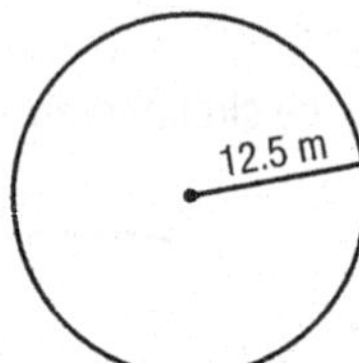

g

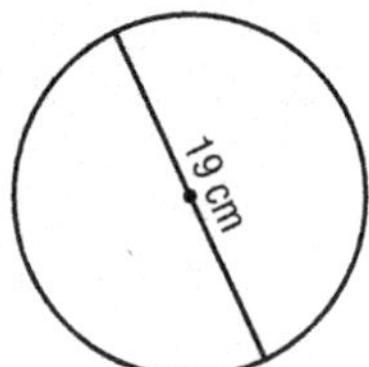

g

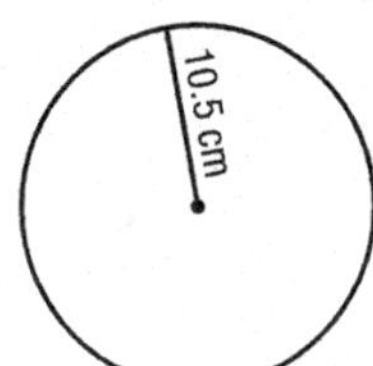

i

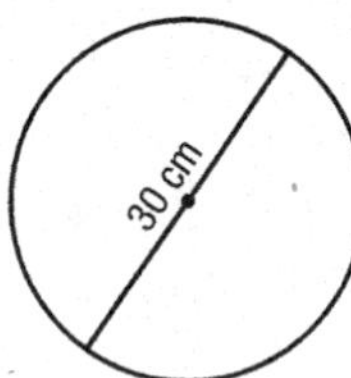

j

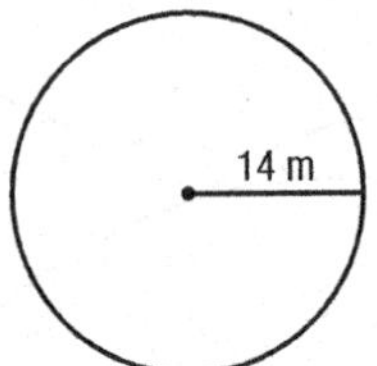

k

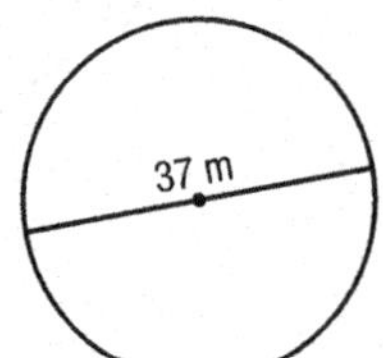

l

3 Calculate the perimeter of each half circle or quarter circle correct to two decimal places.

a

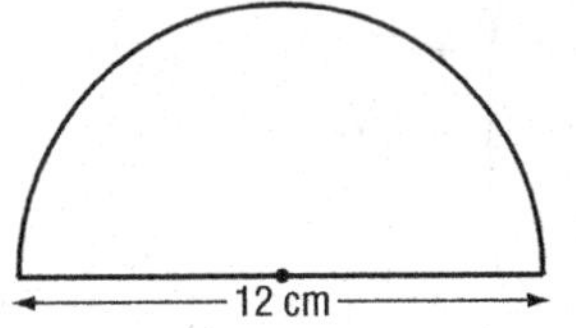

b

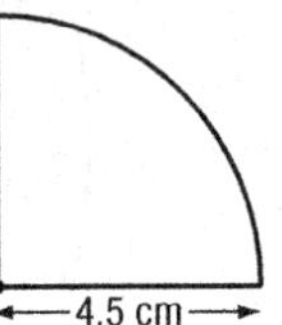

c

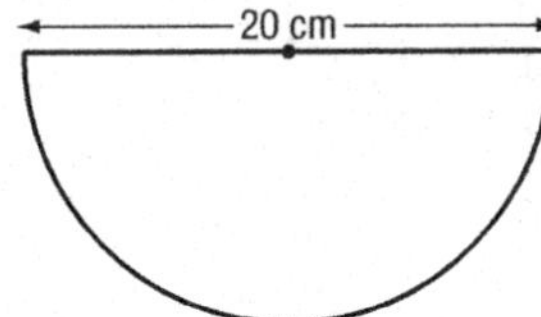

d

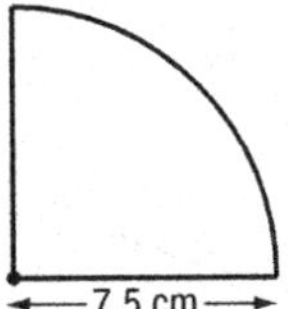

e

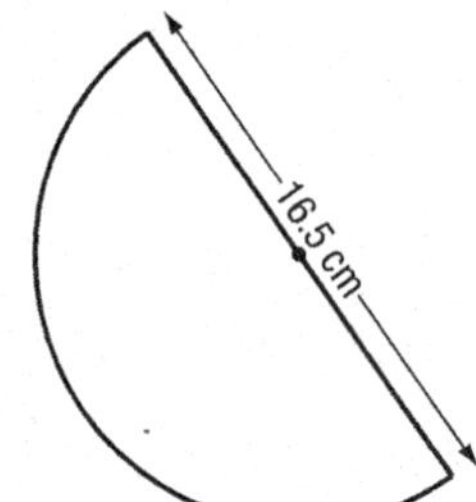

f

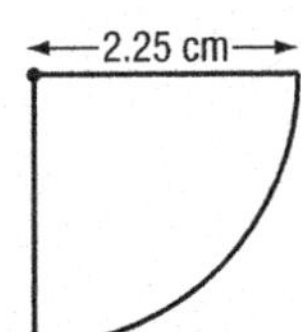

Calculate radius, diameter and circumference of circles

1 Calculate the diameter of circles that have these radii.

a	6.5 cm	**b**	29.5 cm	**c**	13.25 cm	**d**	2.4 m	**e**	11.6 m	**f**	7.8 m
g	38 mm	**h**	106 mm	**i**	23.9 cm	**j**	44.7 cm	**k**	5.85 m	**l**	12.55 m

2 Calculate the radius of circles that have these diameters.

a	36 m	**b**	54 cm	**c**	18.6 m	**d**	7.2 cm	**e**	81.5 mm	**f**	25.8 m
g	14.3 cm	**h**	63 m	**i**	49.1 cm	**j**	135 mm	**k**	74.7 m	**l**	9.3 cm

3 Calculate the circumference of circles that have these diameters.

a	6.25 cm	**b**	8.4 m	**c**	11.75 cm	**d**	9.3 m	**e**	4.75 cm	**f**	15.2 m
g	25.7 cm	**h**	18.6 m	**i**	22.8 cm	**j**	31.9 m	**k**	16.7 cm	**l**	40.65 m

4 Calculate the circumference of circles that have these radii.

a	5.45 m	**b**	12.8 cm	**c**	7.75 m	**d**	11.7 cm	**e**	8.9 m	**f**	10.65 cm
g	15.95 m	**h**	9.6 cm	**i**	4.87 m	**j**	13.56 cm	**k**	2.69 m	**l**	18.74 cm

Remember

If you know the circumference of a circle, you can use the formula $C \div 3.14$ to calculate the diameter. You can then find the radius by dividing the diameter by 2.

An estimate of diameter can be made by dividing the circumference by 3 (3.14 rounded to the nearest whole number).

5 Estimate the diameter of circles that have these circumferences.

a	63 cm	**b**	30 cm	**c**	100 cm	**d**	48 cm	**e**	135 cm	**f**	84 cm
g	150 m	**h**	117 cm	**i**	168 m	**j**	210 m	**k**	124 m	**l**	250 m
m	70 cm	**n**	175 cm	**o**	200 cm	**p**	110 cm	**q**	275 cm	**r**	350 cm

6 Calculate the diameter of circles that have these circumferences, correct to two decimal places.

a	35 cm	**b**	24 cm	**c**	16 cm	**d**	87 cm	**e**	63 cm	**f**	91 cm
g	77 m	**h**	117 m	**i**	142 m	**j**	106 m	**k**	125 m	**l**	139 m
m	252 cm	**n**	186 cm	**o**	211 cm	**p**	170 cm	**q**	209 cm	**r**	155 cm

7 Calculate the radius of circles that have these circumferences, correct to two decimal places.

a	50 cm	**b**	34 cm	**c**	78 cm	**d**	42 cm	**e**	95 cm	**f**	66 cm
g	83 m	**h**	110 m	**i**	154 m	**j**	126 m	**k**	232 m	**l**	105 m
m	218 cm	**n**	179 cm	**o**	300 cm	**p**	266 cm	**q**	148 cm	**r**	325 cm

Solve word problems related to circumference

1 The diameter of a bike tyre is 60 cm.

- **a** Calculate the distance travelled by the tyre in one revolution, correct to one decimal place.
- **b** How far, in metres, will the tyre travel in 50 revolutions?

2 A wheel has a circumference of 1.5 metres. What is its radius, correct to two decimal places?

3 A cylindrical water tank has a base diameter of 4.5 metres. What is the circumference of its base, correct to two decimal places?

4 To keep fit, Obi runs around a circular sports oval each day.

- **a** If the radius of the oval is 37.5 metres, how many metres would Obi run in one lap of the oval? Give your answer correct to one decimal place.
- **b** If Obi runs five laps of the oval each day, how many metres would he run each day?
- **c** After seven days, how far would Obi have run? Give your answer in kilometres, correct to two decimal places.

5 Yerema decides to make a circular garden.

- **a** If she makes the radius of the garden 275 centimetres, how many metres will the diameter be?
- **b** How much fencing will Yerema need to buy to go all the way around the edge of her garden? Give your answer to the nearest tenth of a metre.
- **c** If fencing costs K4.50 per metre, how much will it cost Yerema to fence her garden?

6 A hotel constructs a circular swimming pool with a diameter of 25 metres.

- **a** What is the circumference of the pool?
- **b** If a fence is to be built two metres from the pool edge all the way around the pool, what will be the circumference of the fence?
- **c** If pool fencing costs K70 per metre, how much will the fence cost?

7 Kathy makes circular tablecloths to sell at the market.

- **a** If the radius of a table is 60 cm and the tablecloth has to hang 20 cm over the edge of the table, what will the diameter of the tablecloth have to be?
- **b** What will be the circumference of the tablecloth? Give your answer to the nearest metre.

8 A tyre completes 596 revolutions to travel one kilometre.

- **a** Calculate the distance in metres travelled by the tyre in one revolution, correct to two decimal places.
- **b** What is the radius of the tyre in centimetres, correct to two decimal places?

8.2.11 Investigate properties of interior and exterior angles of polygons

Find missing interior angles in triangles and quadrilaterals

Calculate the missing angle in each shape below.

1

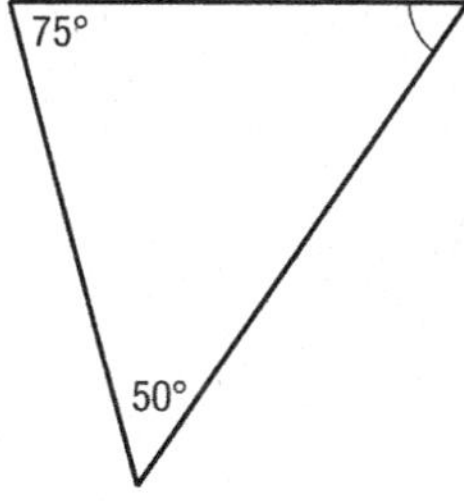

2

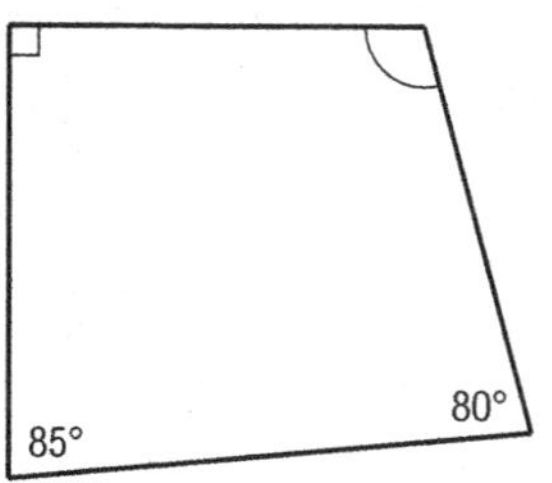

3

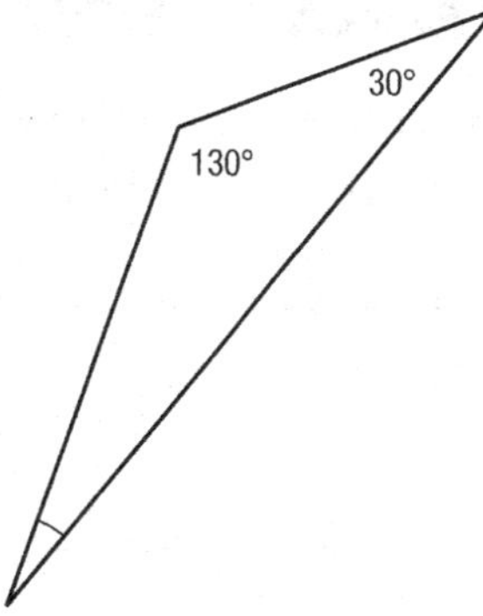

4

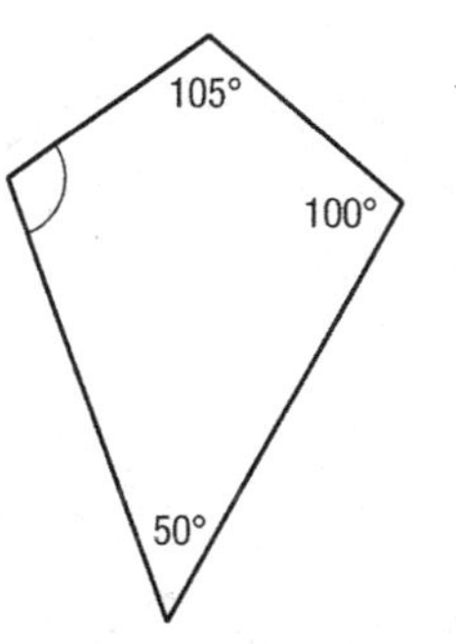

5

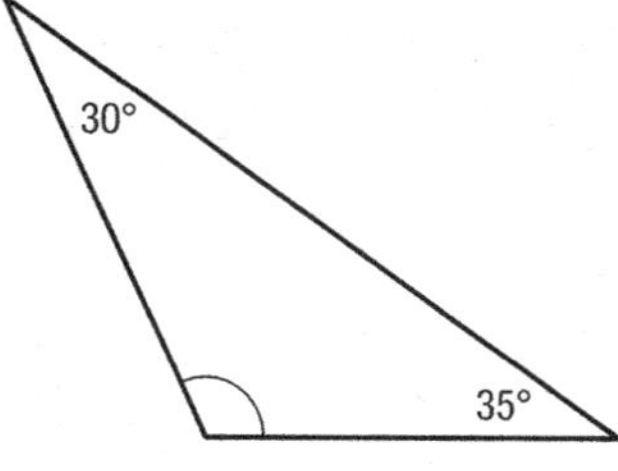

6

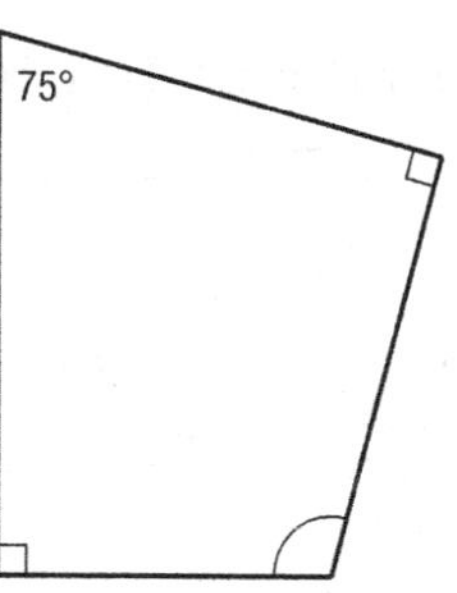

7

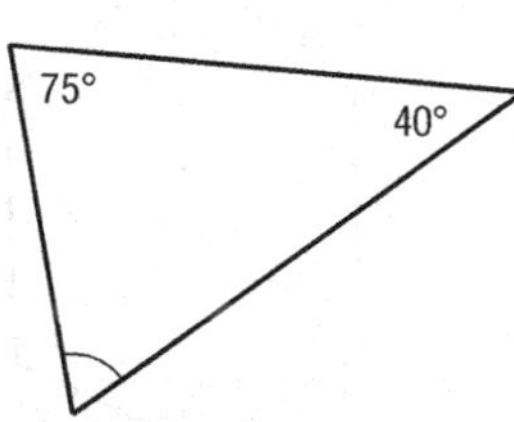

8

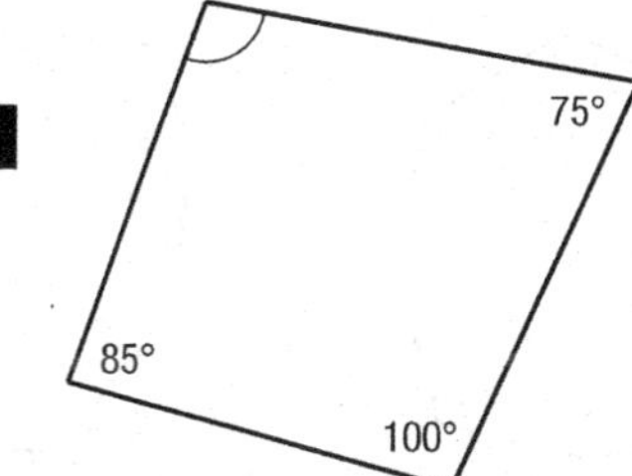

9

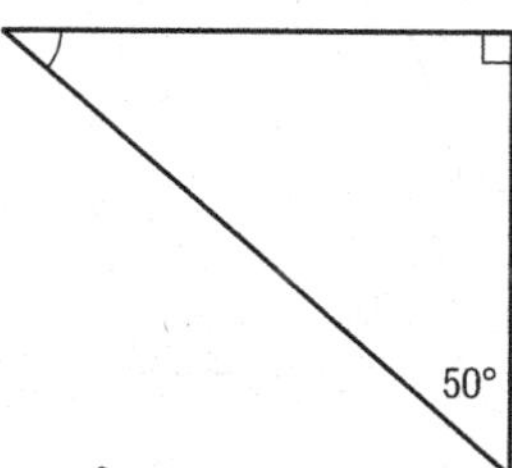

10

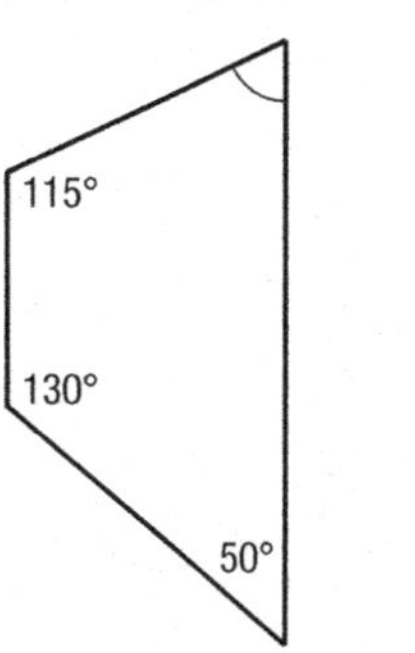

11

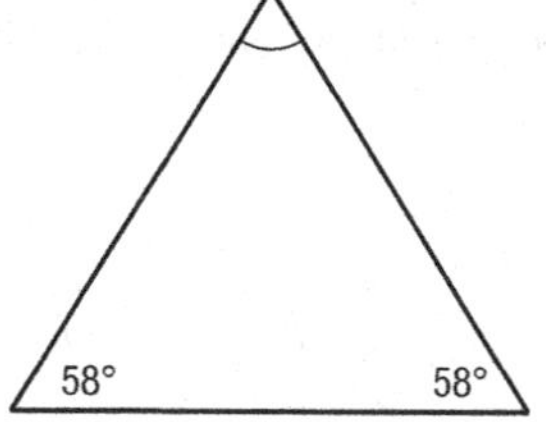

12

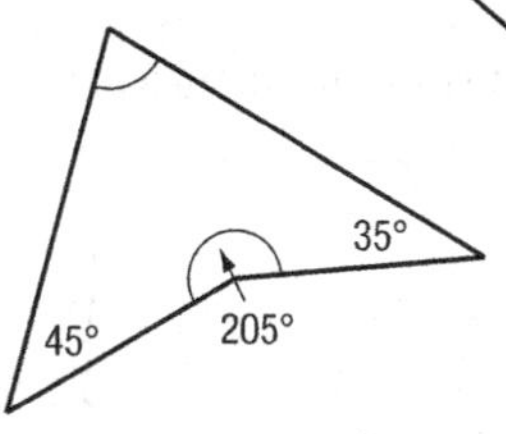

13

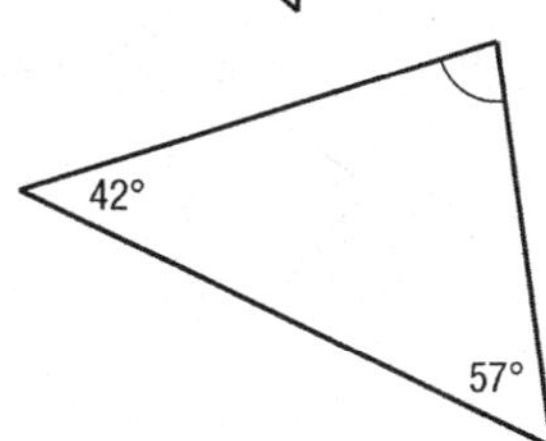

14

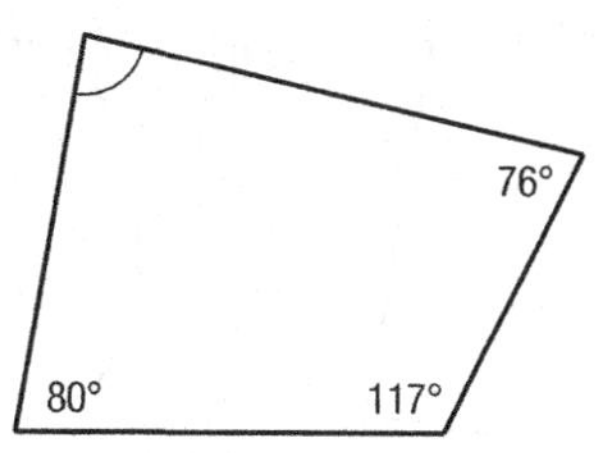

15

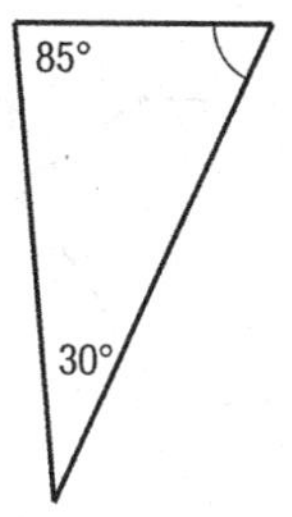

Calculate the sum of interior angles of polygons

Help Box

There are two formulae that can be used to find the sum of the interior angles of polygons.

1. Number of internal triangles × 180°

   The polygon shown here has 3 internal triangles so the sum of its internal angles is $3 \times 180° = 540°$.

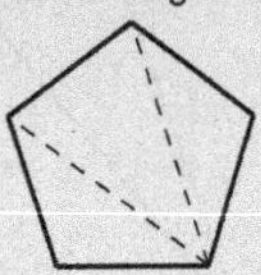

2. $(n - 2) \times 180°$ where n is the number of sides on the polygon.

   The polygon shown here has 4 sides so $(4 - 2) \times 180° = 2 \times 180° = 360°$.

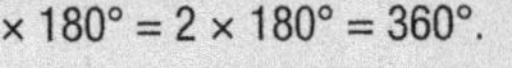

1 Use the first formula to calculate the sum of the interior angles of each polygon.

a 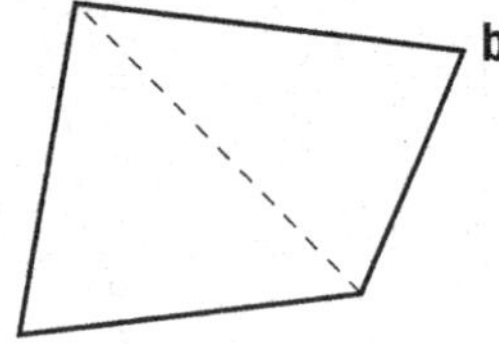

b 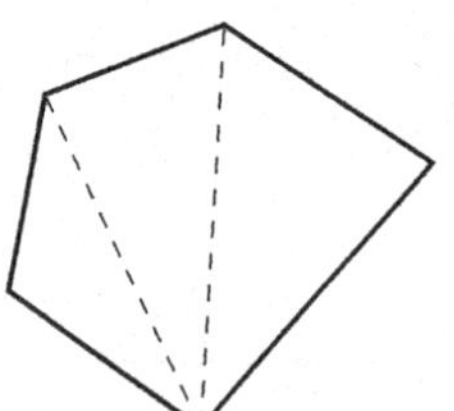

c 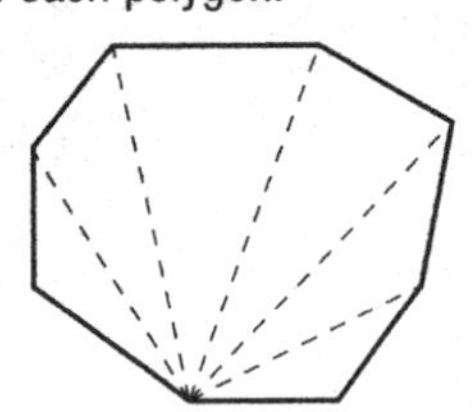

d 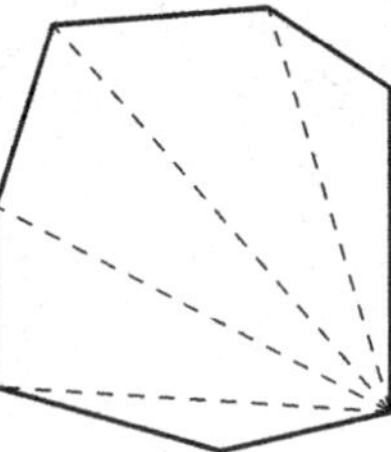

e 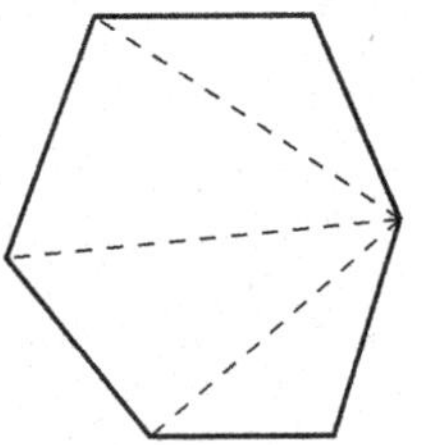

f 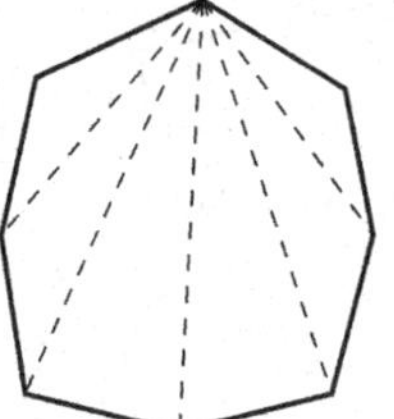

g 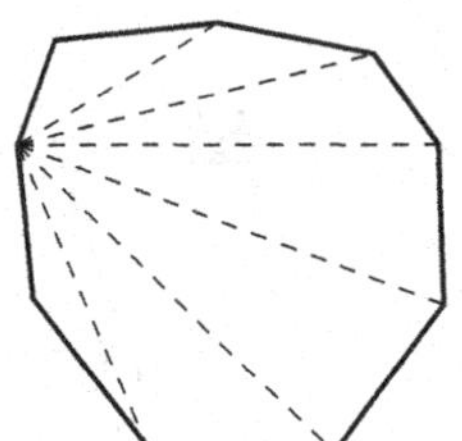

h 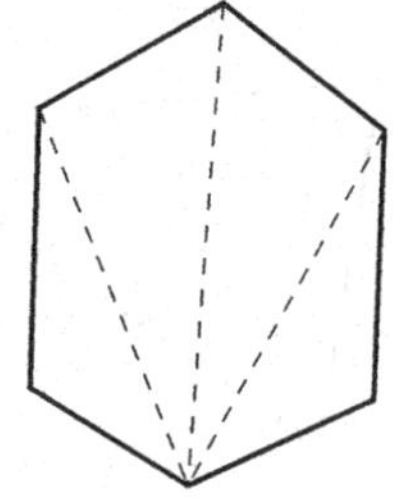

2 Use the second formula to calculate the sum of the interior angles of each polygon.

a 

b 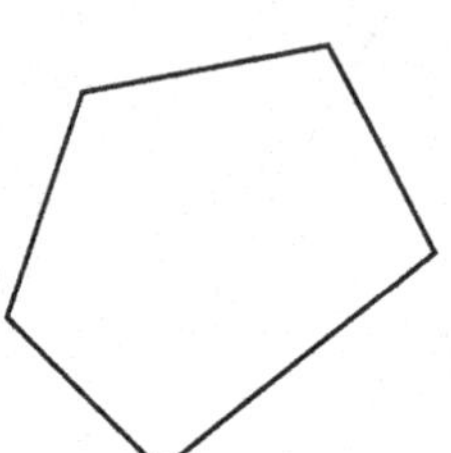

c 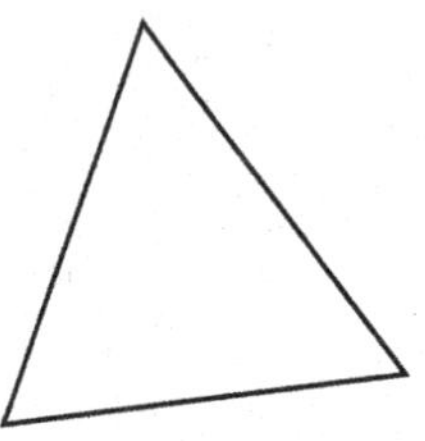

d 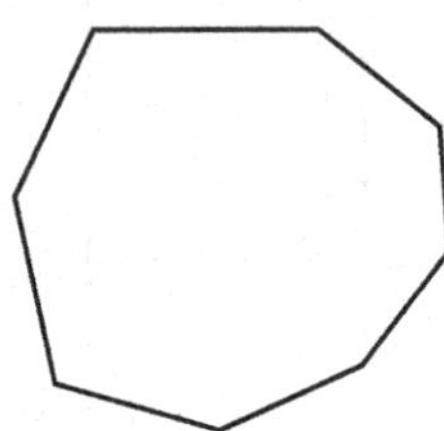

e 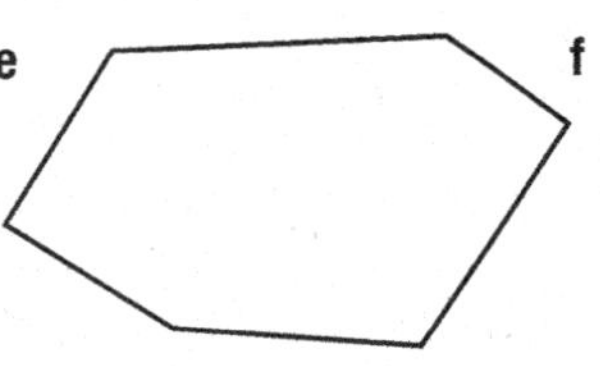

f

g 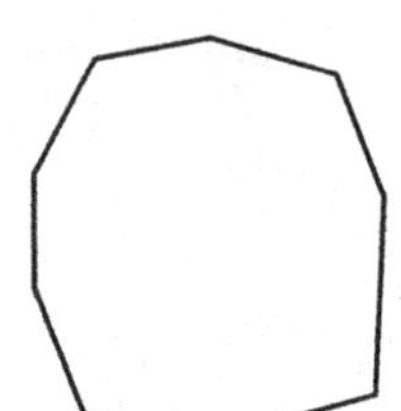

h 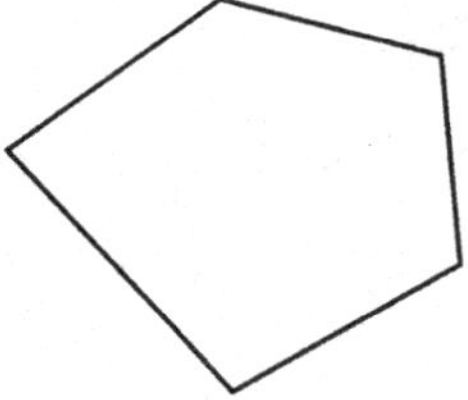

Find the value of unknown angles in polygons

Use the formula $(n-2) \times 180°$ to find the angle sum of each polygon and then calculate the value of the unknown angle in each polygon.

1

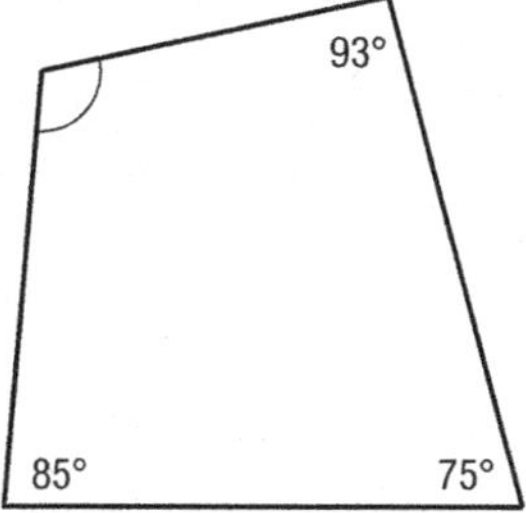

2

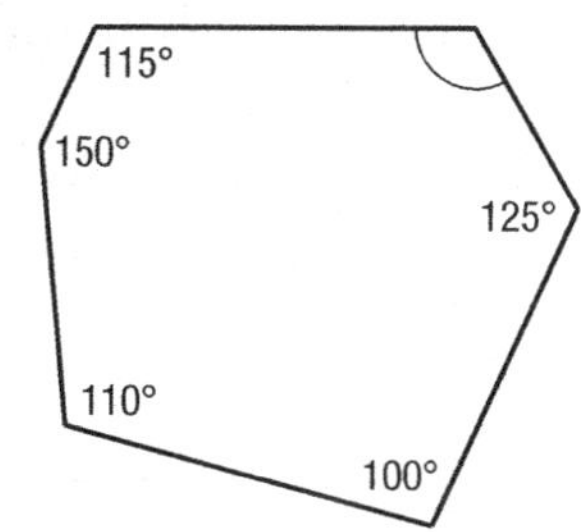

3

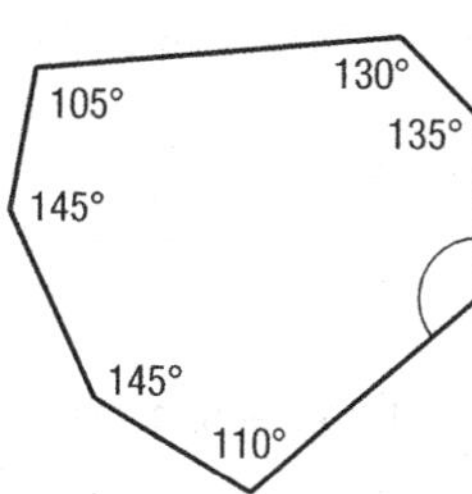

4

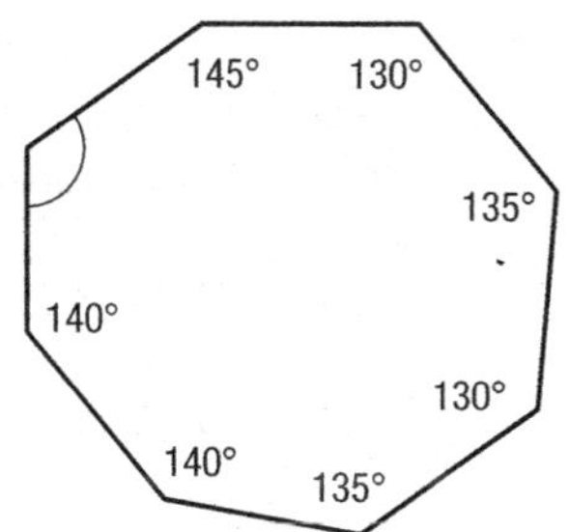

5

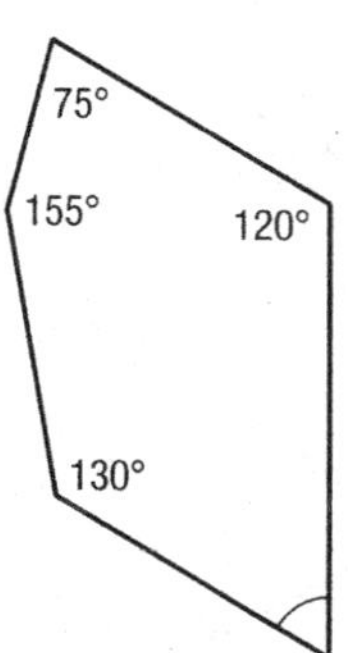

6

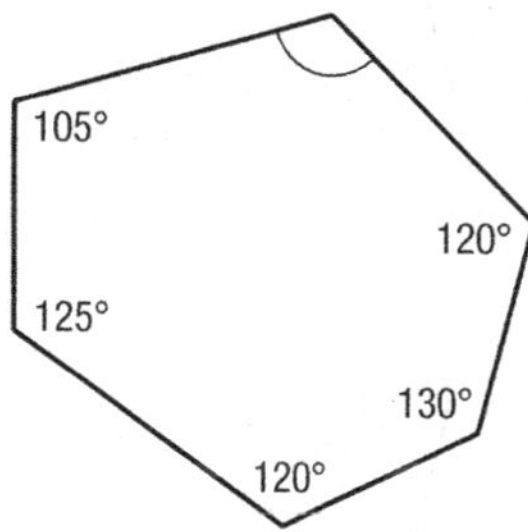

7

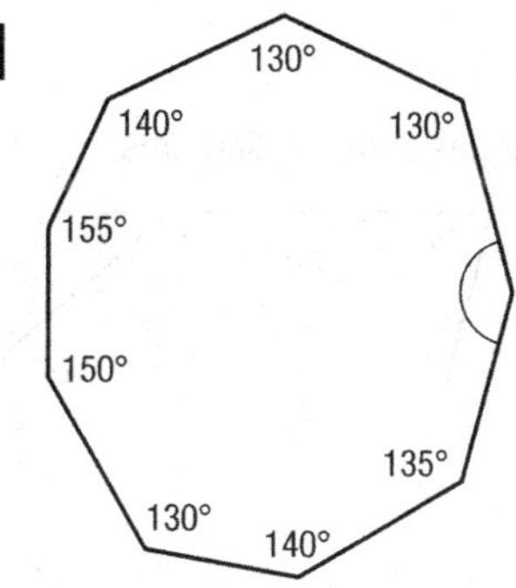

8

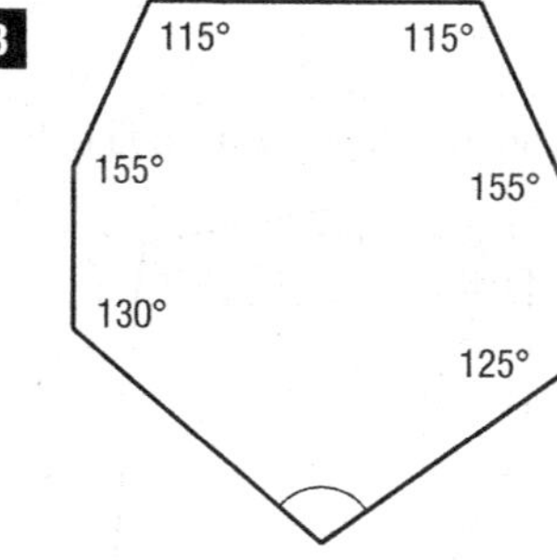

9

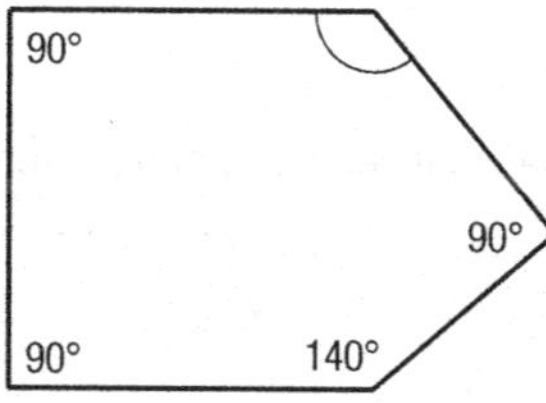

10

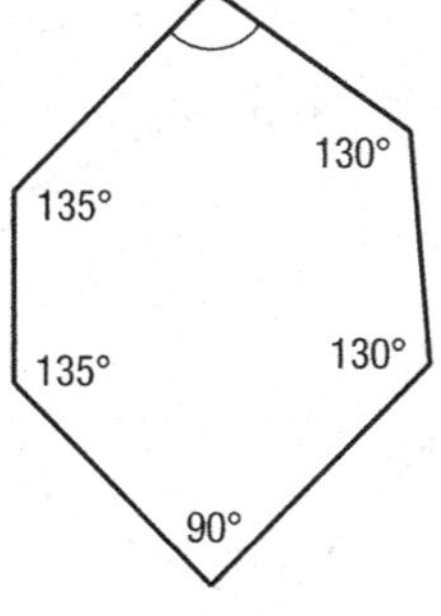

11

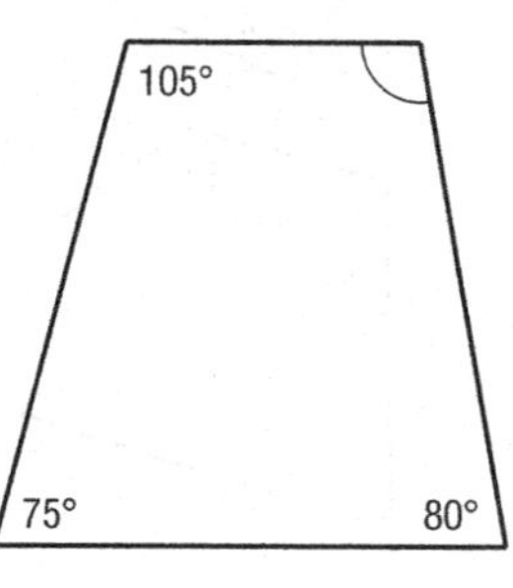

12

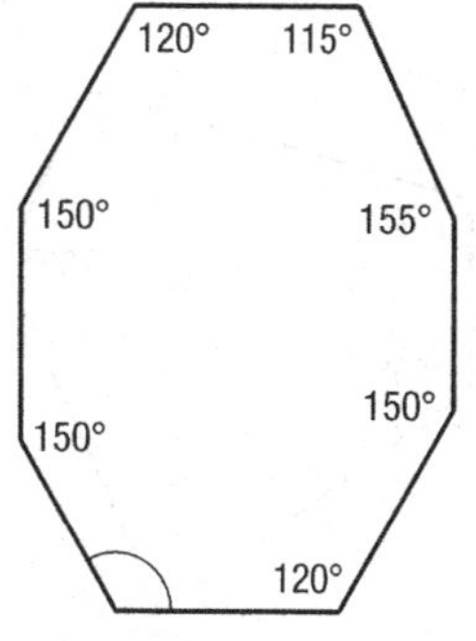

Calculate interior and exterior angles on polygons

Help Box

The sum of two interior angles of a triangle is equal to the exterior angle opposite them.

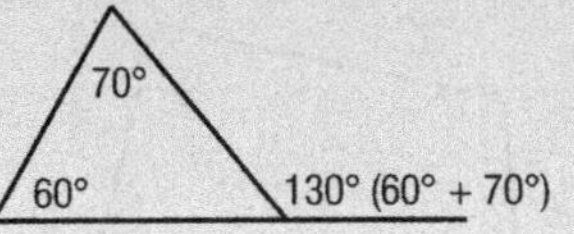

1 Calculate the exterior angle in each diagram.

a 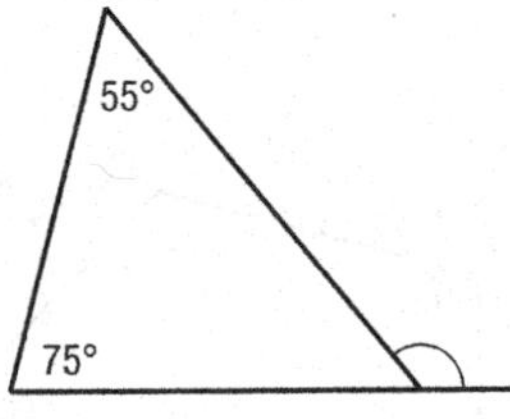

b 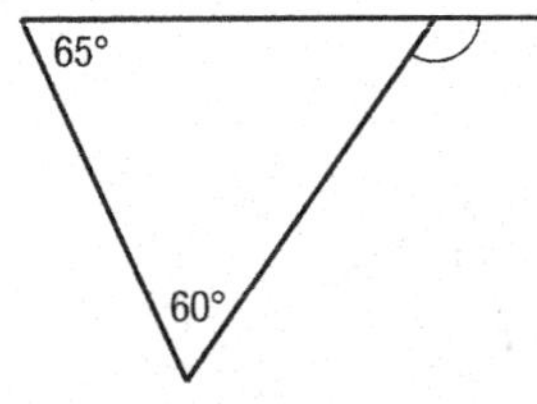

c 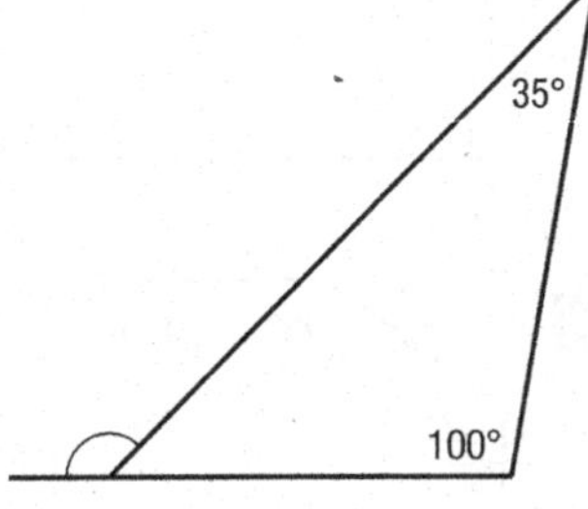

d 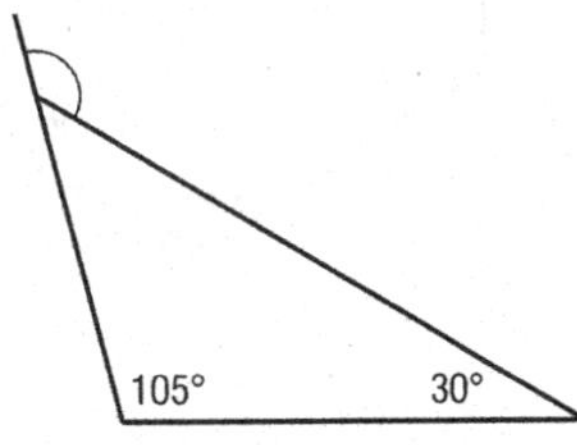

e 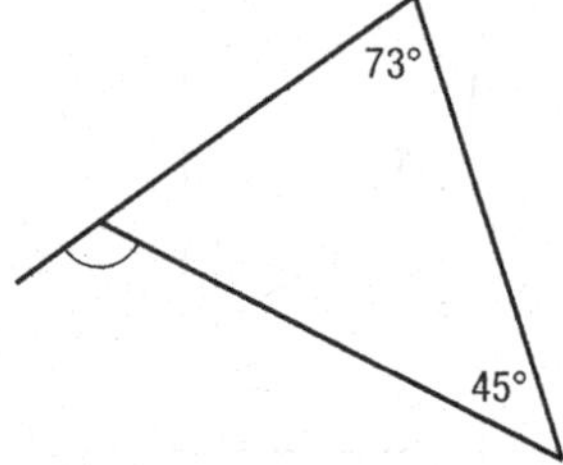

f 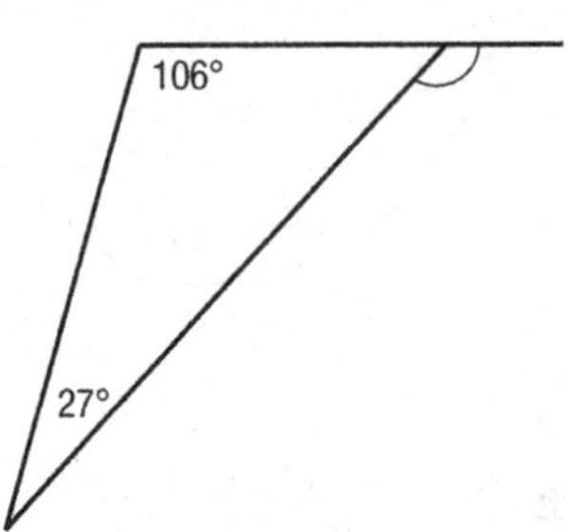

2 Use what you know about interior and exterior angles on polygons to find the missing angles in these diagrams.

a 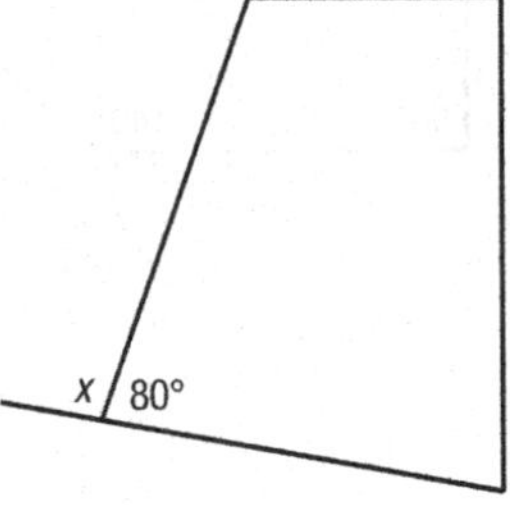

b 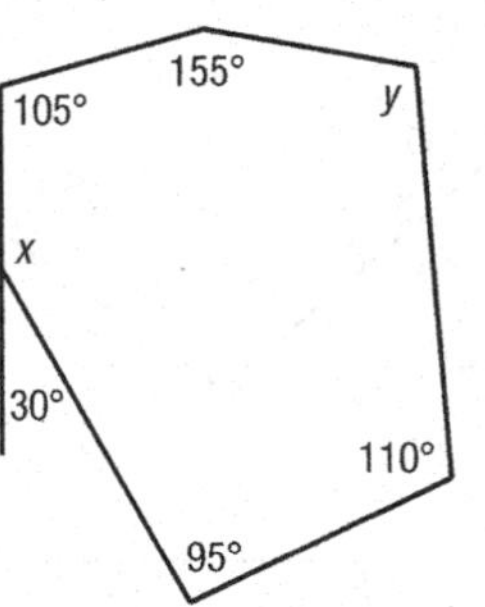

c 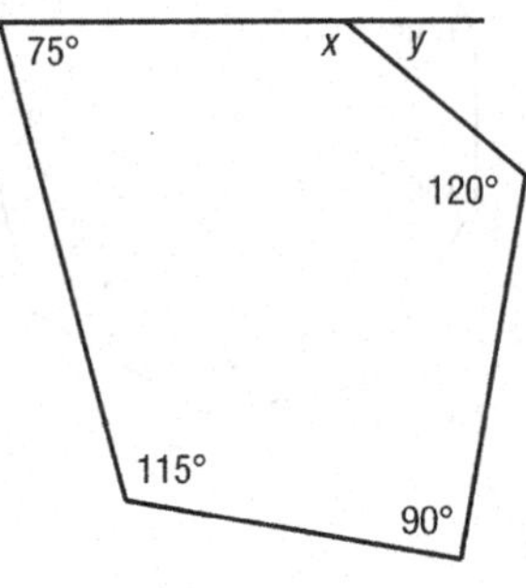

d 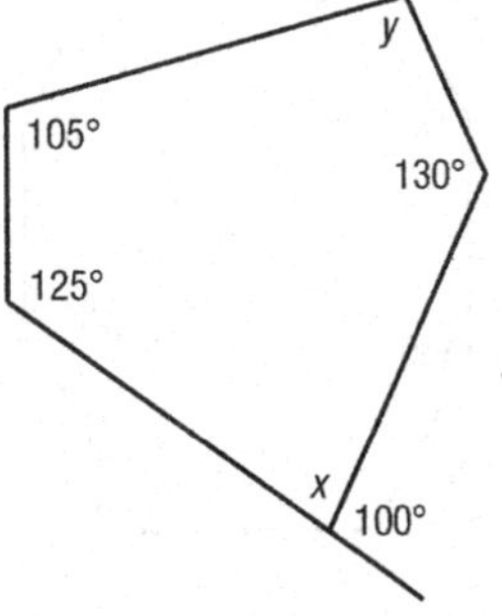

e 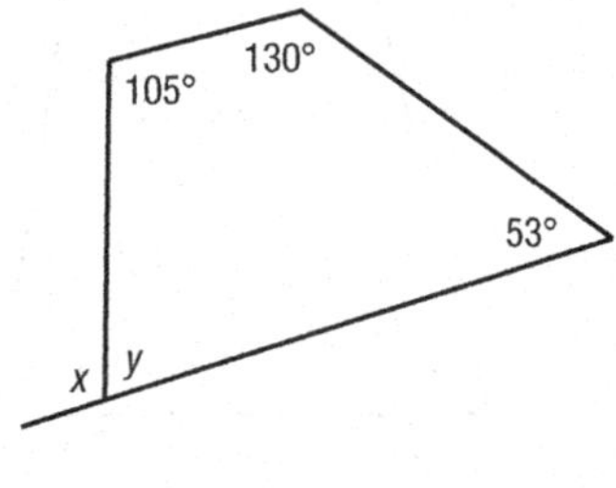

f 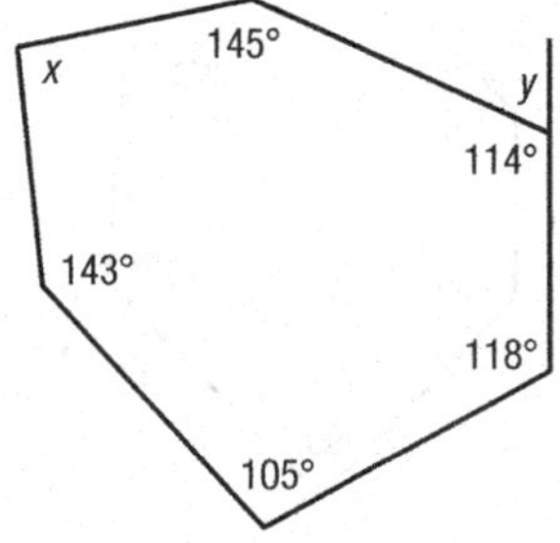

8.2.12 Use appropriate terms to describe angles and shapes accurately

Calculate complementary and supplementary angles

Help Box

When two adjacent angles add to 90° they are called complementary angles.

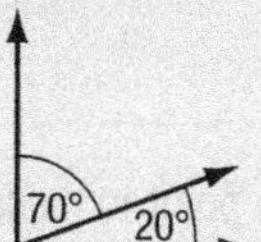

When two adjacent angles add to 180° they are called supplementary angles.

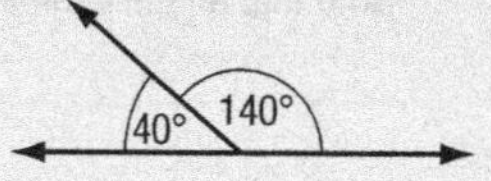

1 Calculate the size of the missing angle in each diagram.

a

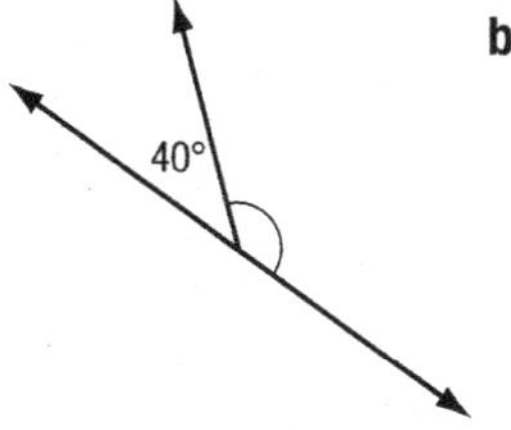

b

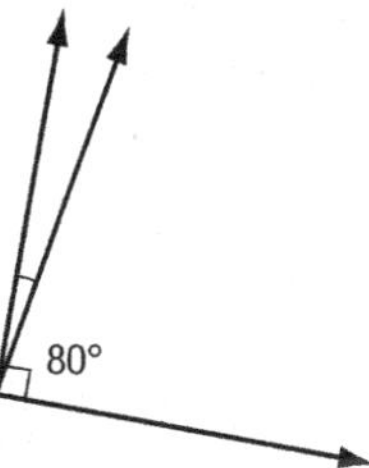

c

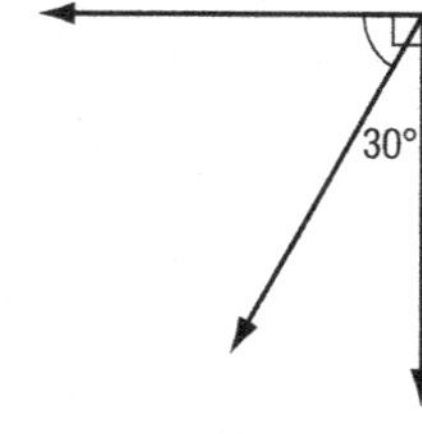

d

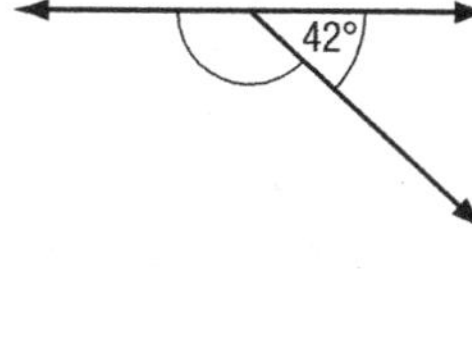

e

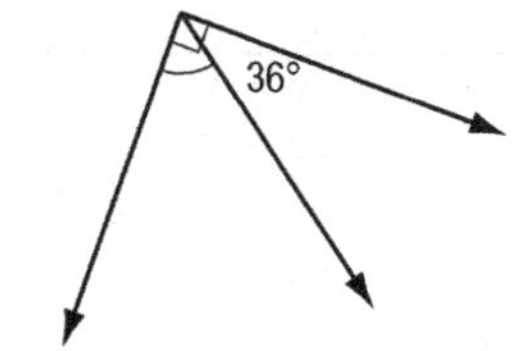

f

106°

g

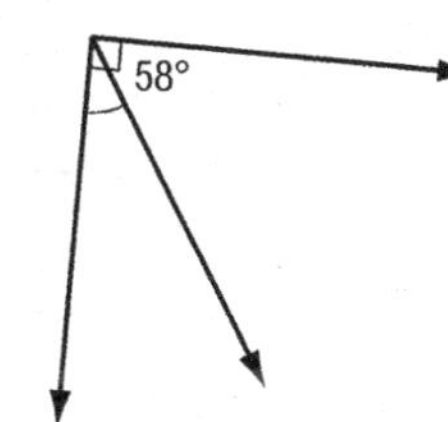

h

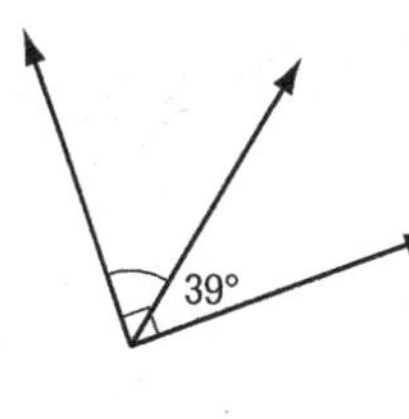

i

27°

j

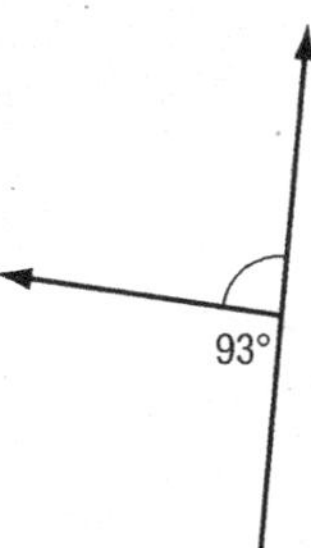

k

l

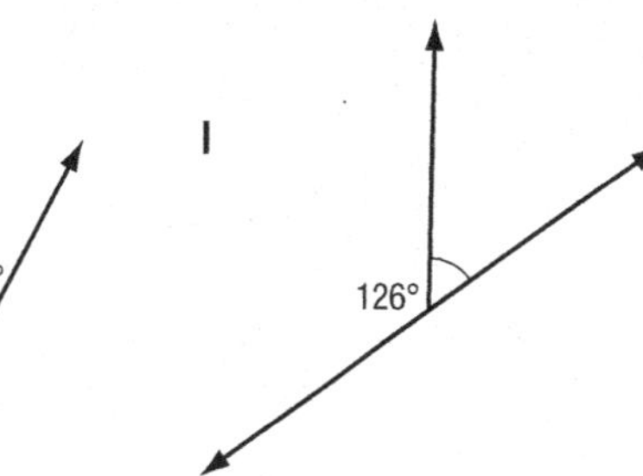

2 Write the complementary angle to each of these angles.

a	66°	b	37°	c	18°	d	59°	e	21°	f	42°
g	54°	h	9°	i	26°	j	72°	k	63°	l	11°

3 Write the supplementary angle to each of these angles.

a	84°	b	139°	c	53°	d	109°	e	47°	f	116°
g	12°	h	98°	i	71°	j	36°	k	128°	l	143°

Calculate vertically opposite angles

Help Box

Vertically opposite angles are formed when two straight lines cross each other (intersect).
Two pairs of equal supplementary angles are formed. The vertically opposite angles are equal.

Note that all the angles around the point add up to 360°.

Use what you know about vertically opposite angles and angles around a point to calculate the missing angles in these diagrams.

1

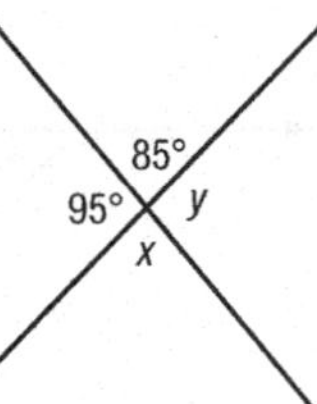

2

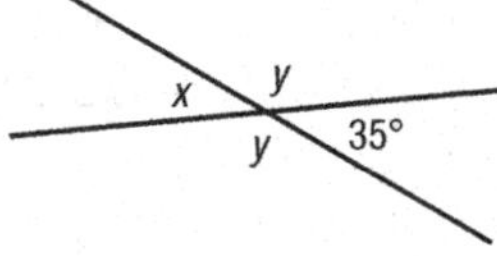

3

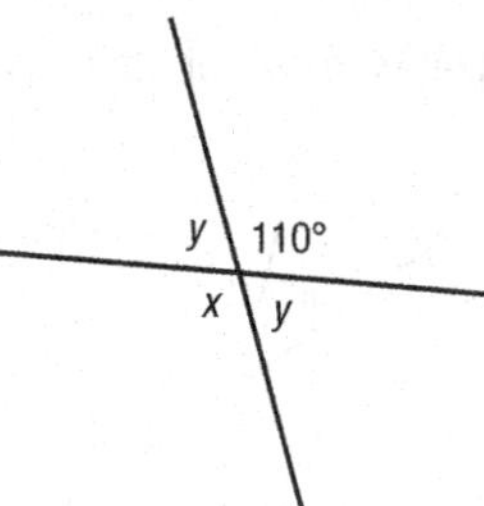

4

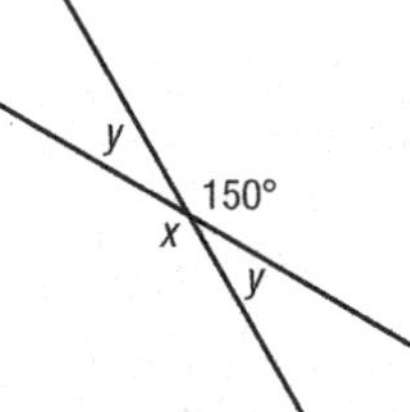

5

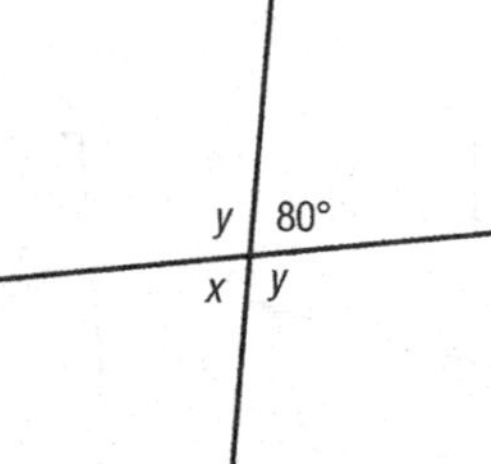

6

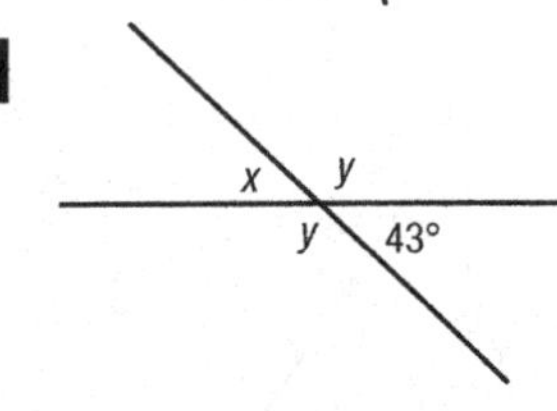

7

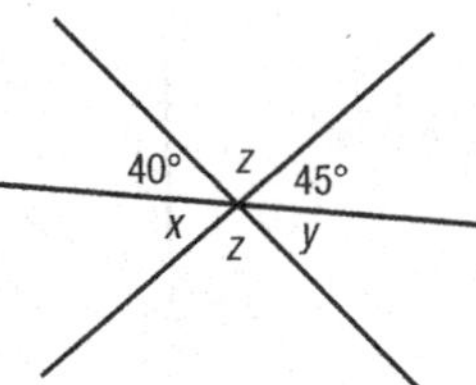

8

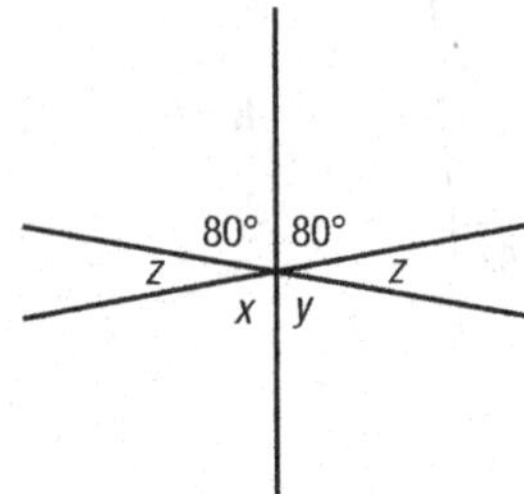

9

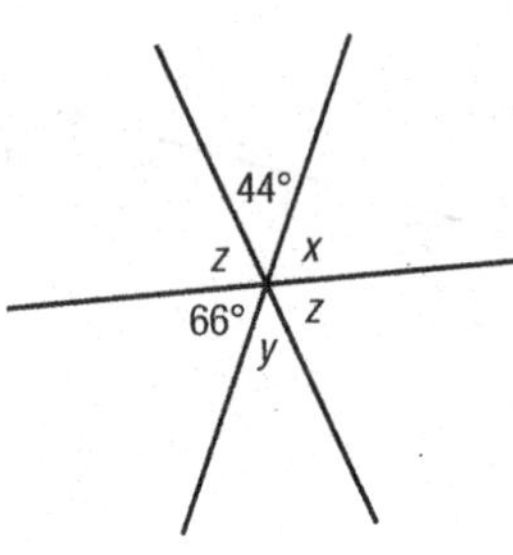

10

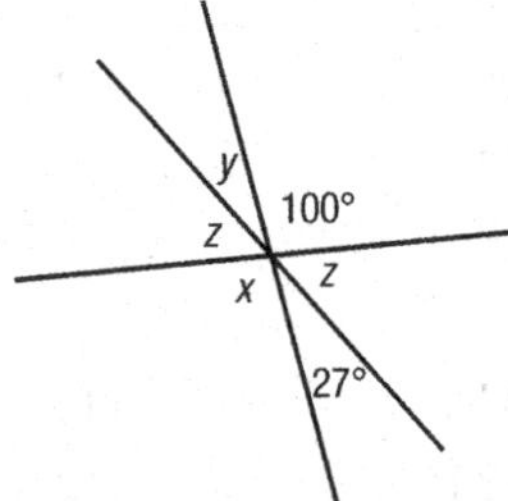

11

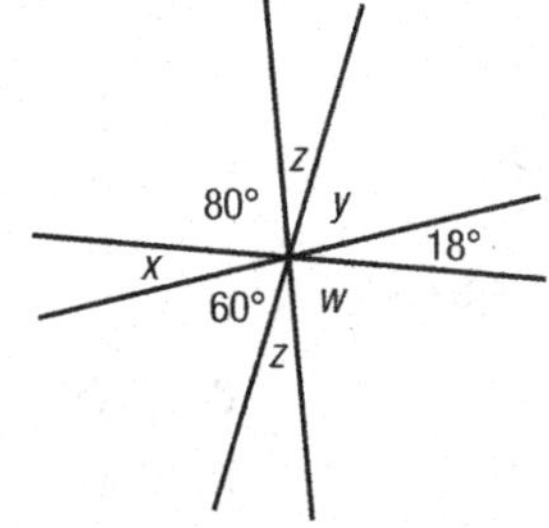

12

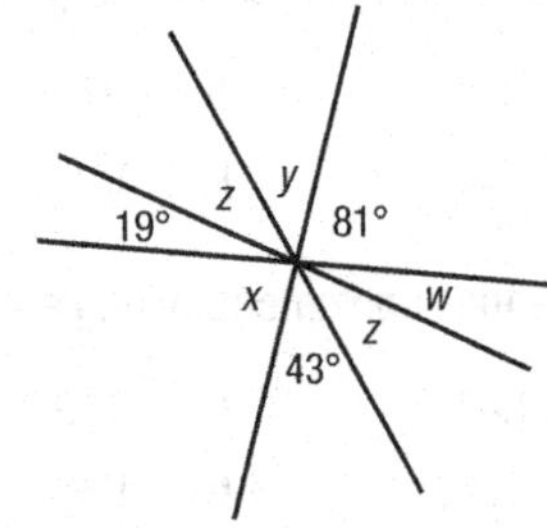

Calculate corresponding, alternate and co-interior angles on parallel lines

Help Box

Three special types of angles are formed when parallel lines are crossed by a transversal.

1. Corresponding angles

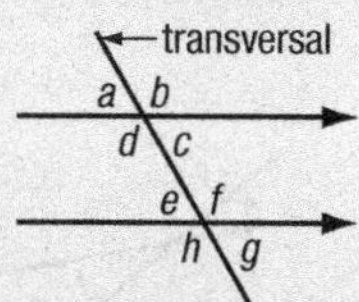

Corresponding angles will be equal if the lines are parallel:
$a = e$, $b = f$, $c = g$, $d = h$

2. Alternate angles

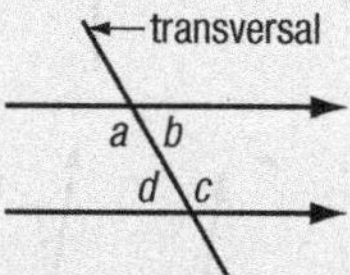

Alternate angles will be equal if the lines are parallel:
$a = c$, $b = d$

3. Co-interior angles

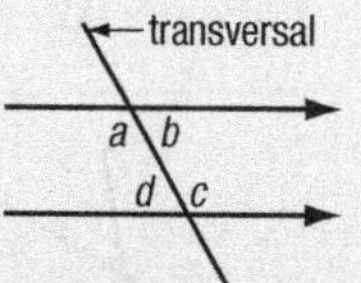

Co-interior angles will total 180° if the lines are parallel:
$a + d = 180°$, $b + c = 180°$

1 Which angle is corresponding to the angle marked D?

a

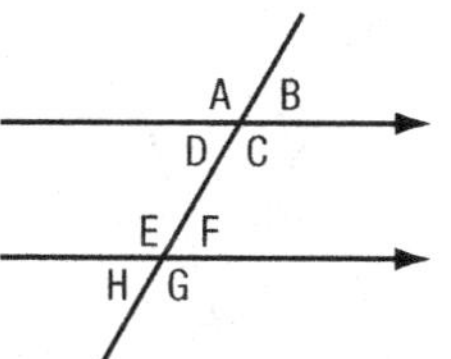

b

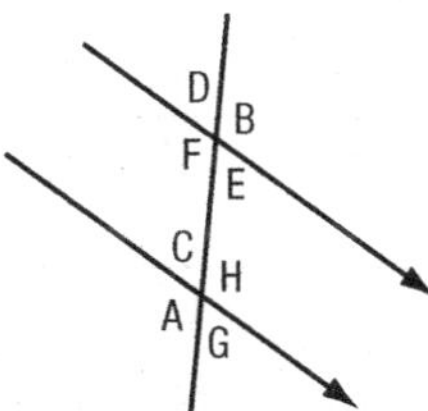

c

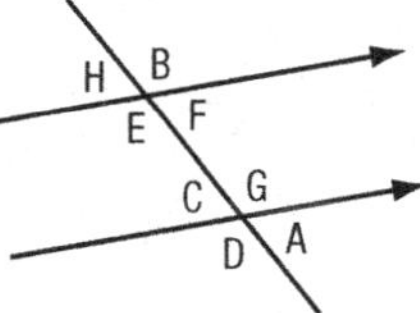

d

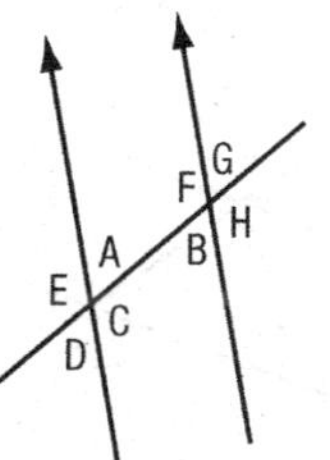

e

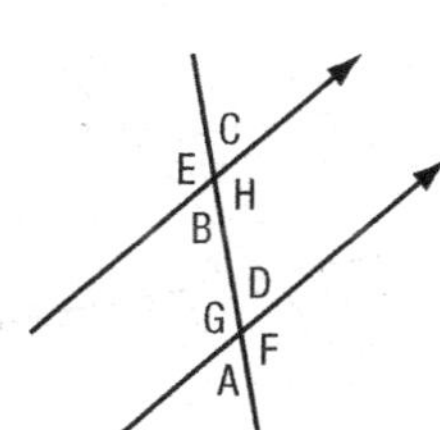

f

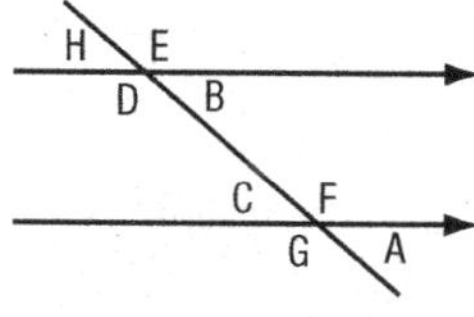

2 Which angle is alternate to the angle marked X?

a

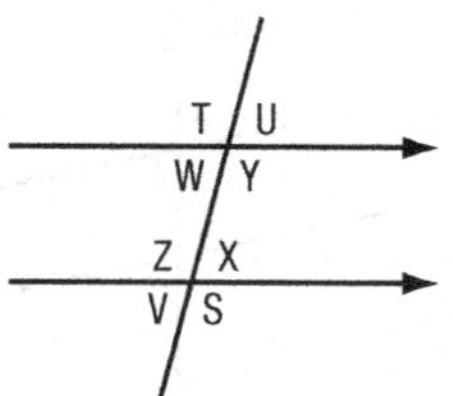

b

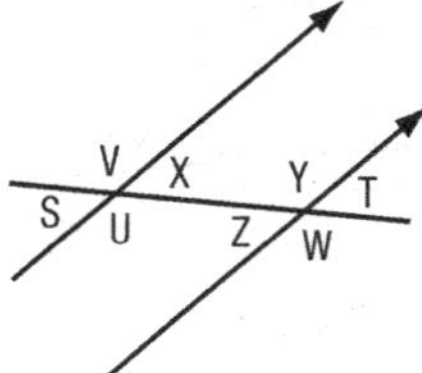

c

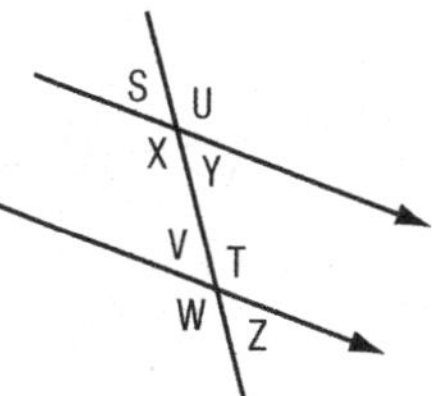

3 Which angle is co-interior to the angle marked B?

a

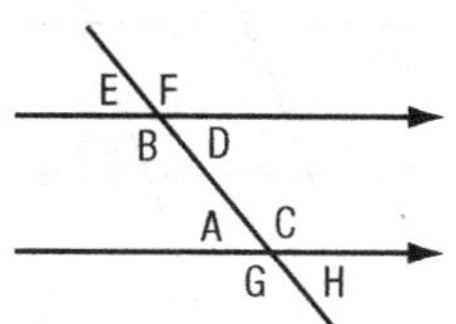

b

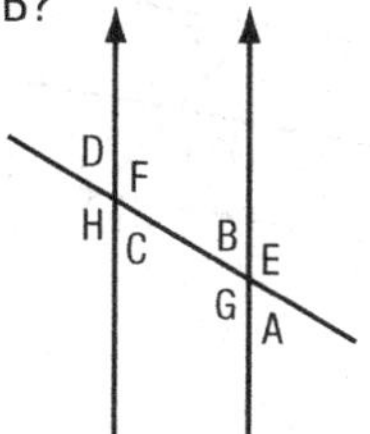

c

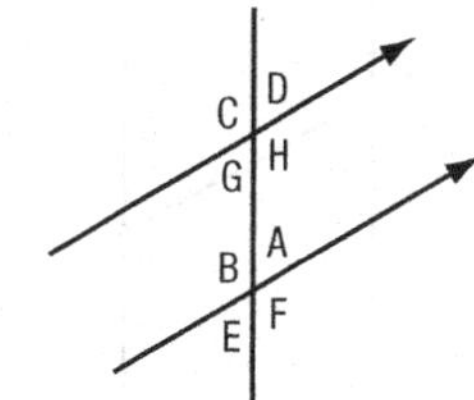

4 Calculate the value of the pronumerals in each diagram.

a

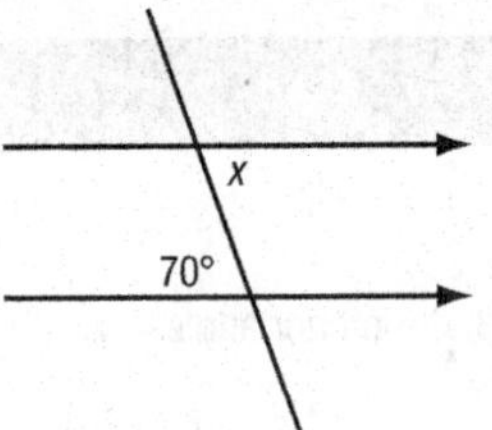

b

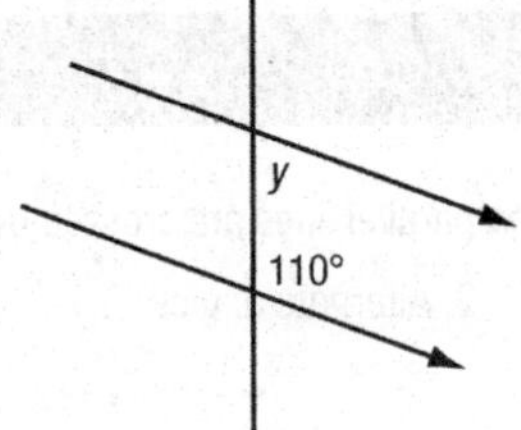

c

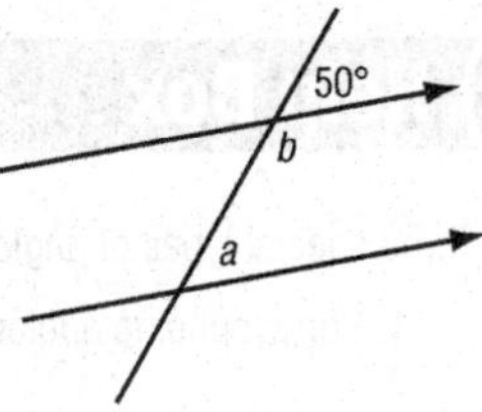

d

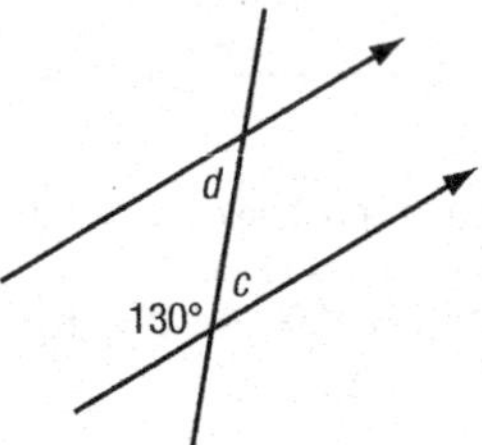

e

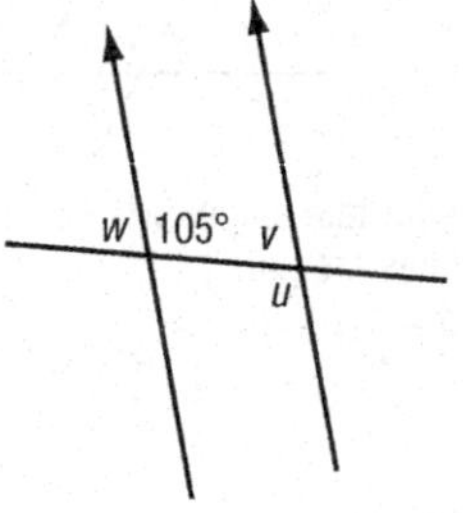

f

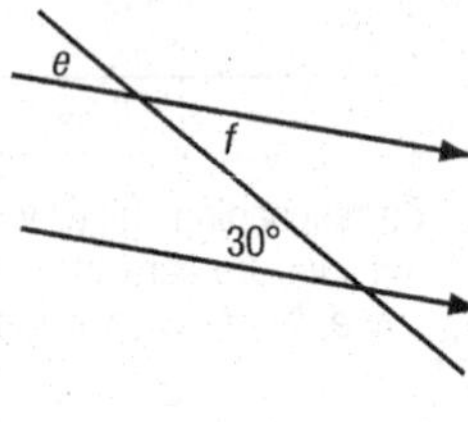

g

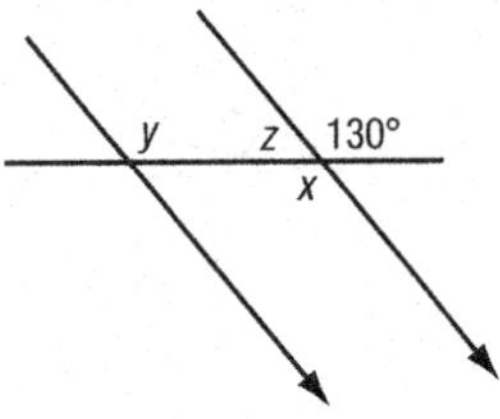

h

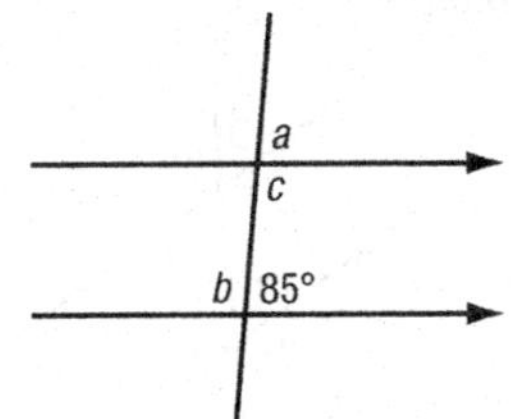

i

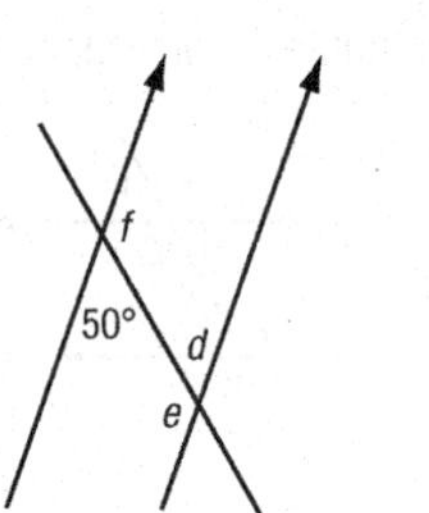

j

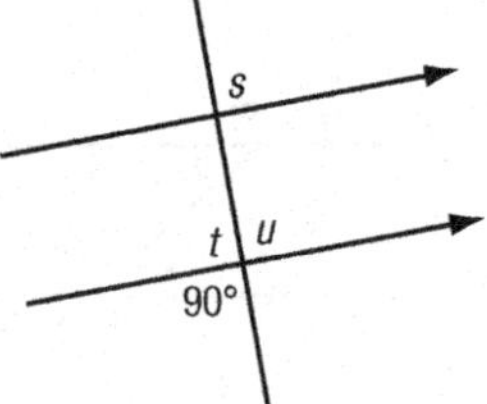

k

l

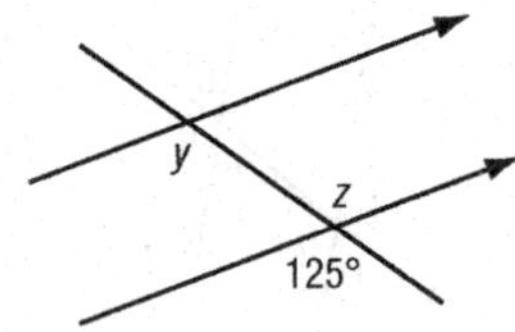

m

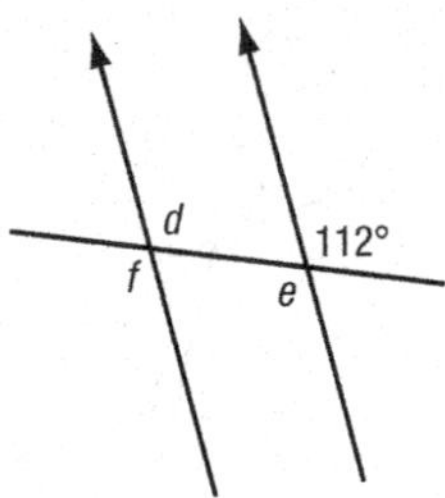

n

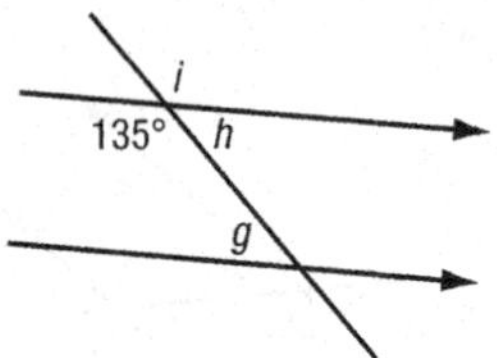

o

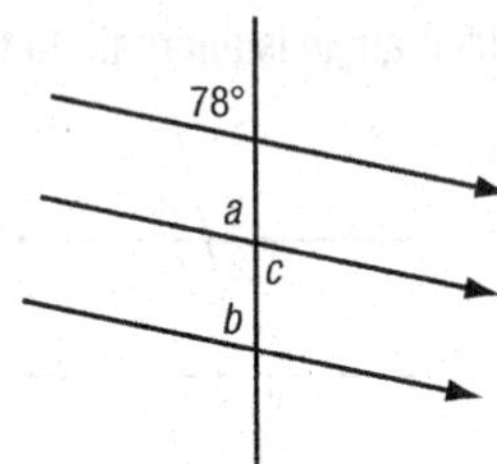

p

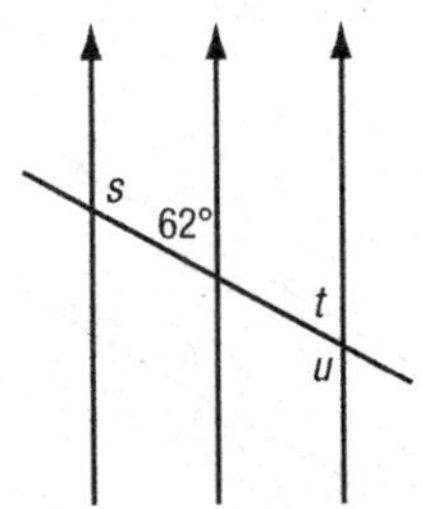

q

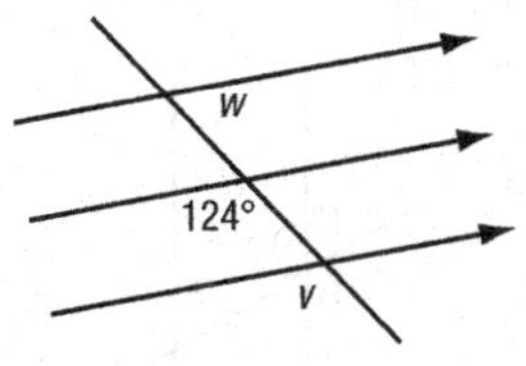

r

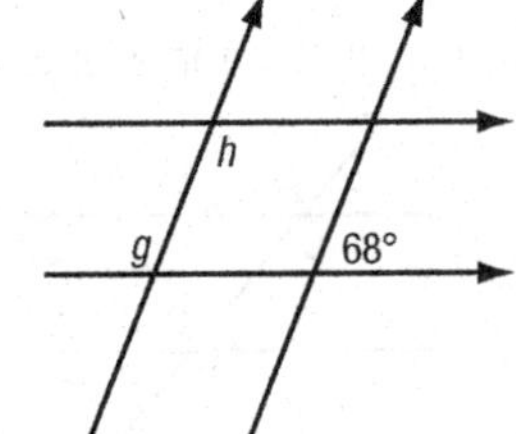

Determine unknown angles

Use what you know about angles to determine the value of the pronumerals in each diagram.

1

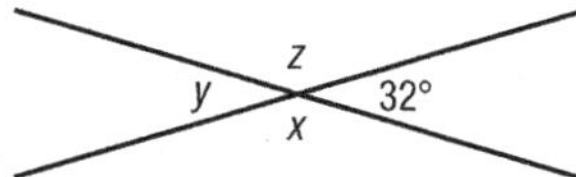

2

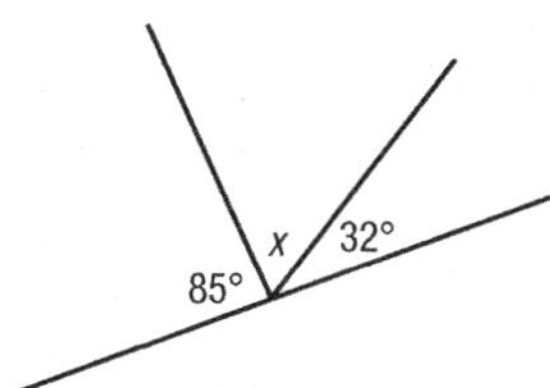

3

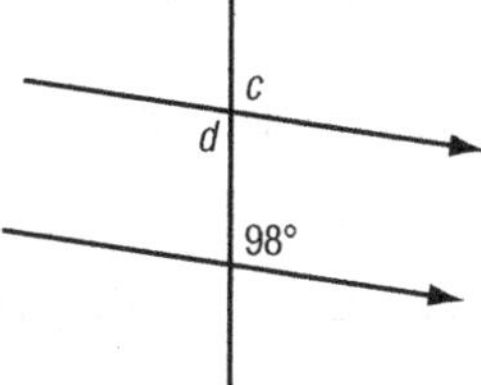

4

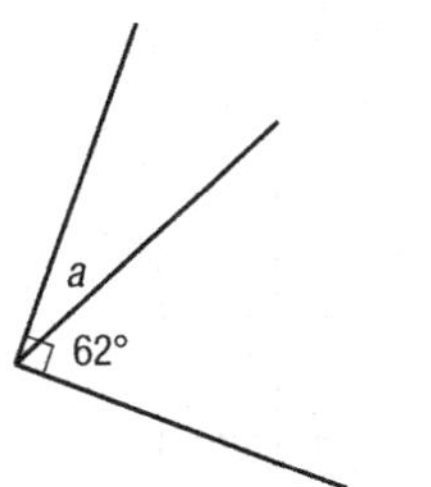

5

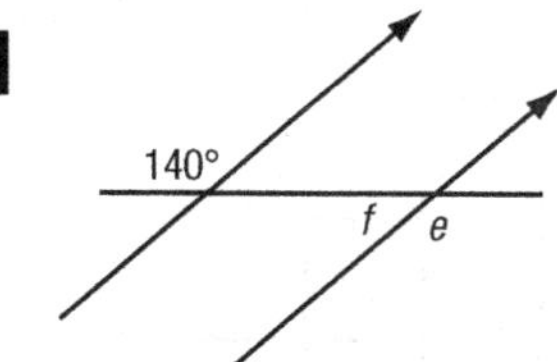

6

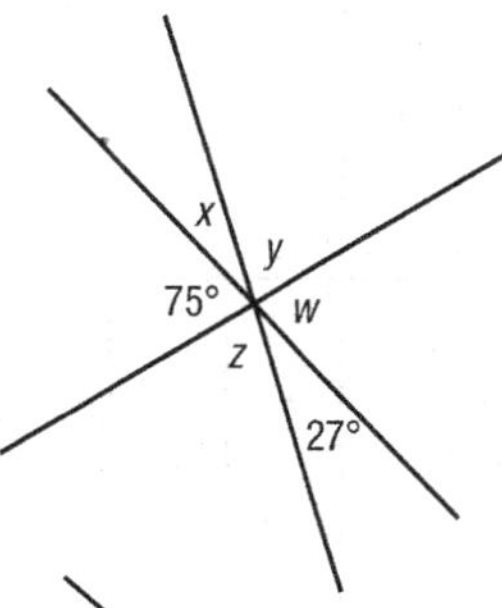

7

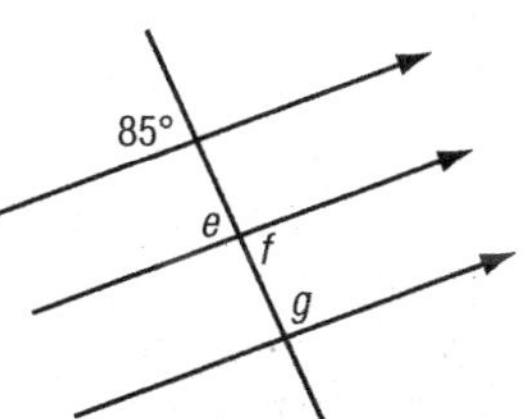

8

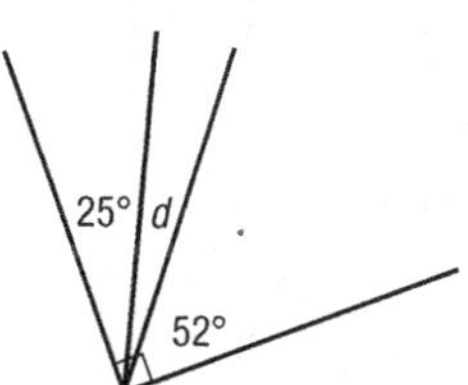

9

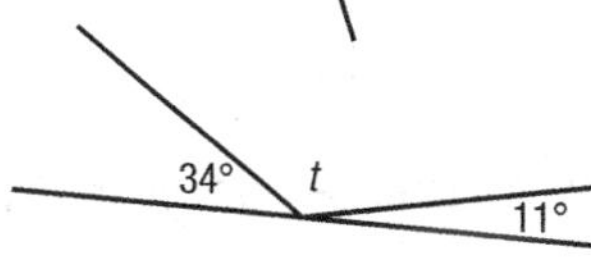

10

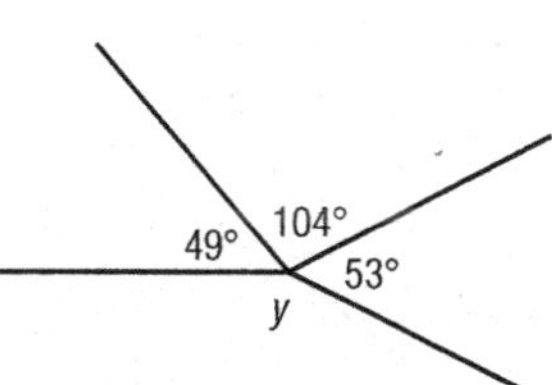

11

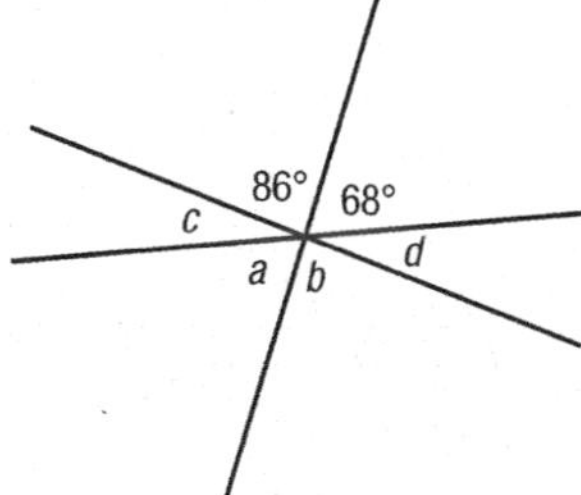

12

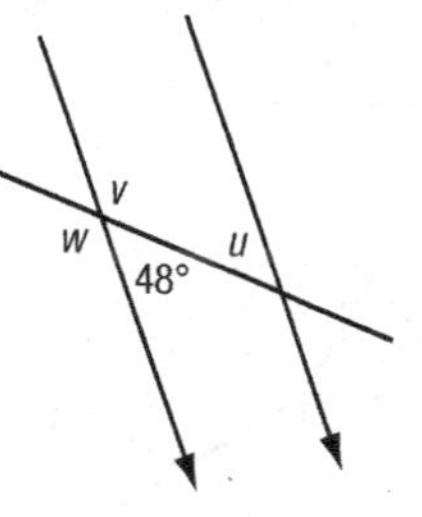

13

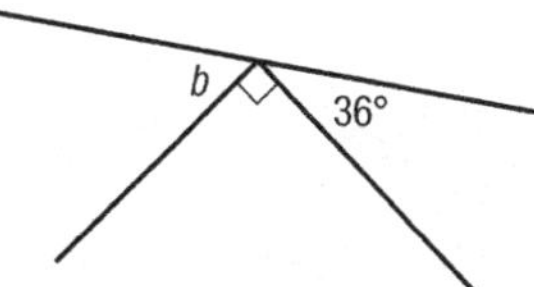

14

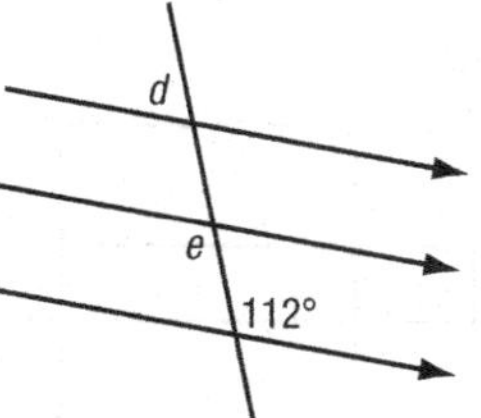

15

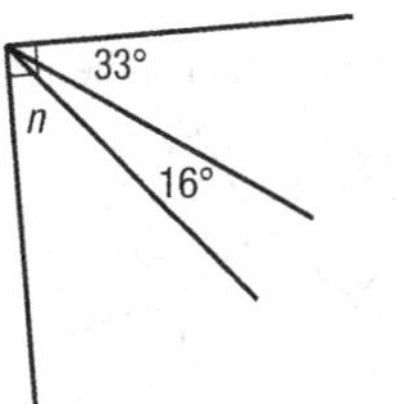

16

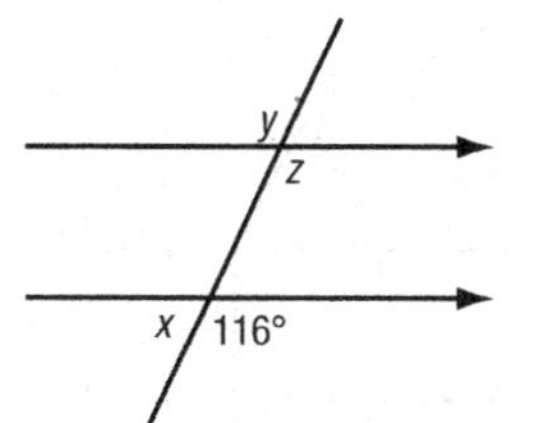

17

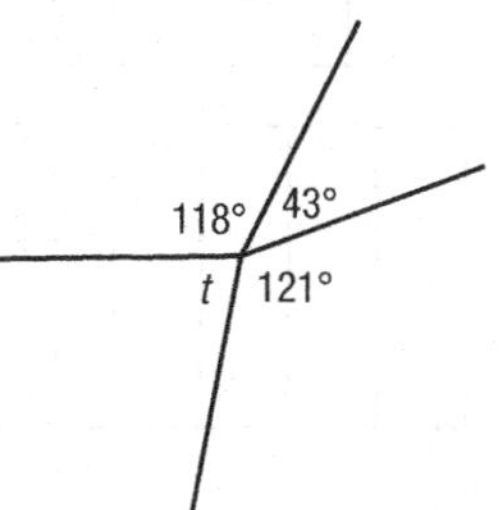

18

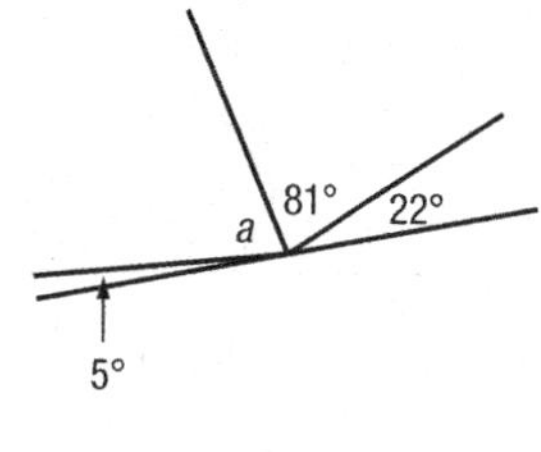

8.2.13 Associate nets with the solids they form

Identify nets for various solids

1 Which of these nets can be folded to form a cube?

a 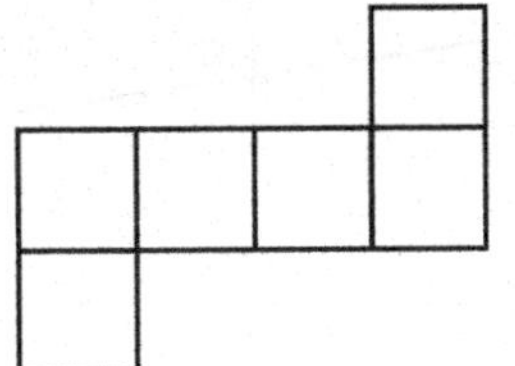

b 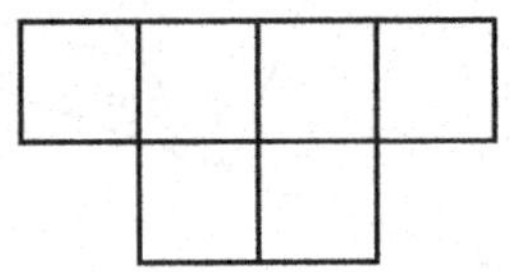

c 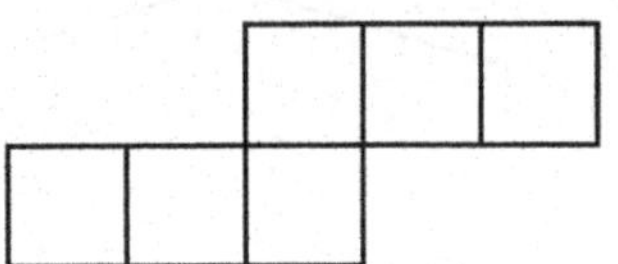

d 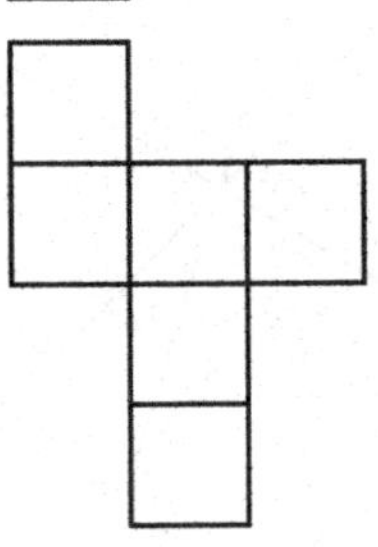

e 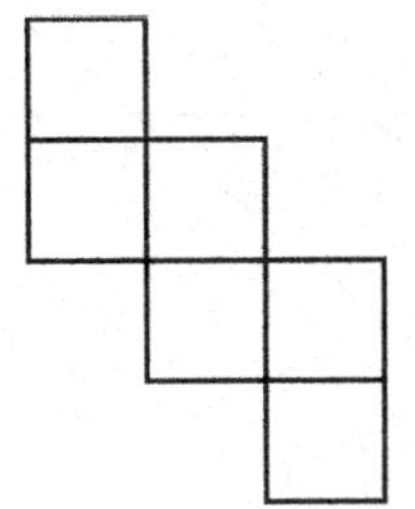

f 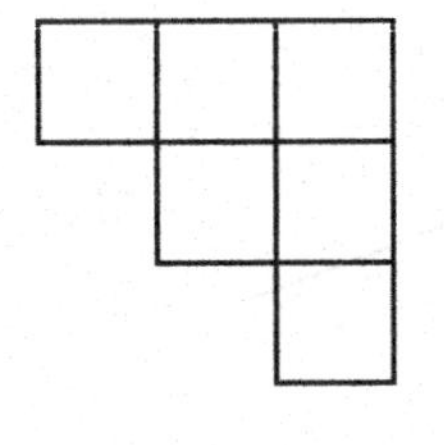

2 Which of these nets can be folded to form a pyramid?

a 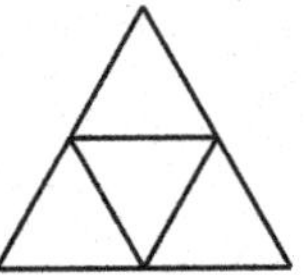

b 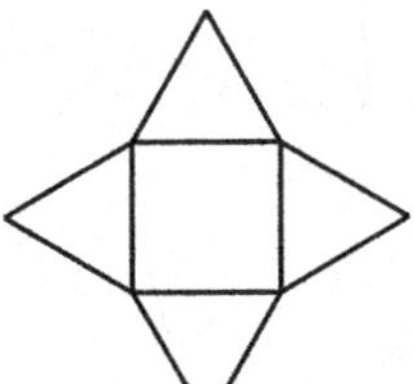

c 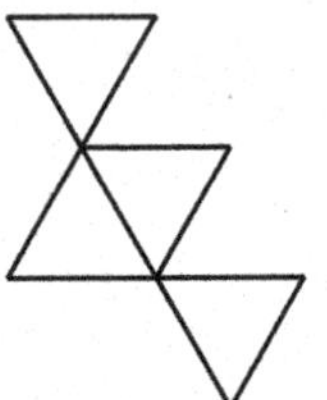

d 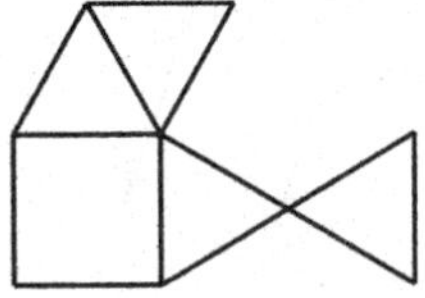

e 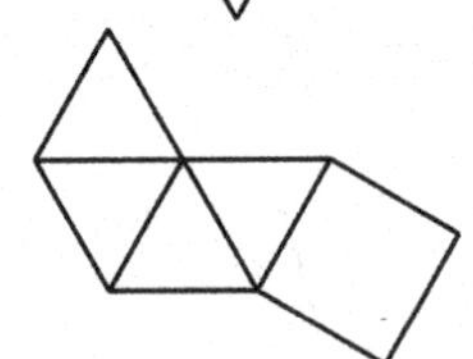

f 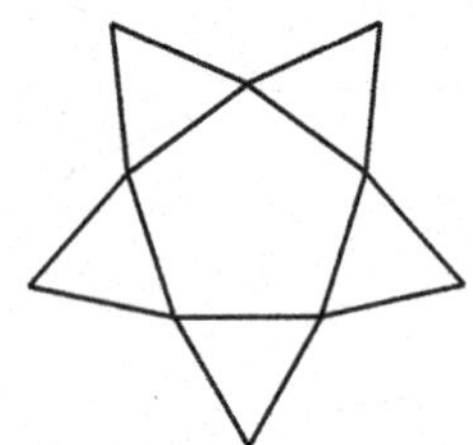

3 Which of these nets can be folded to form a prism?

a 

b 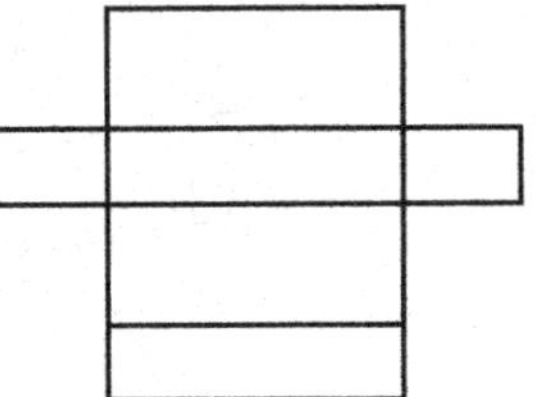

c 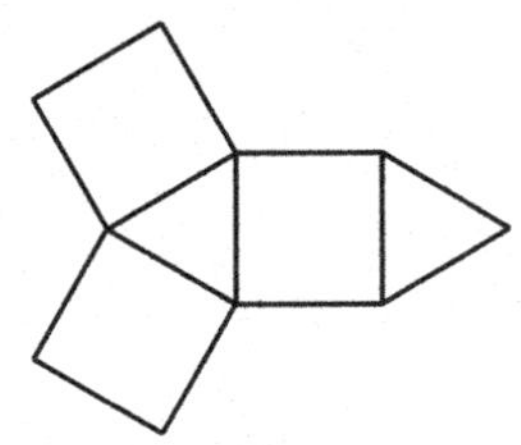

d 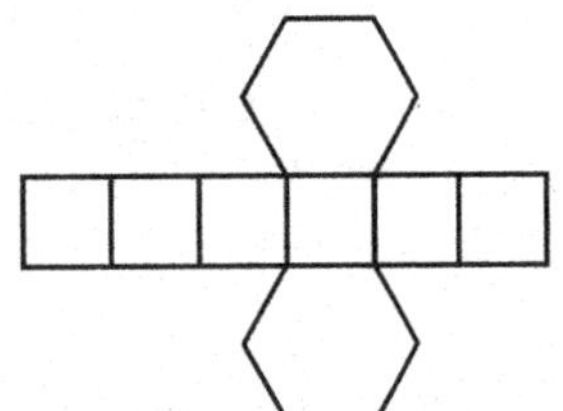

e 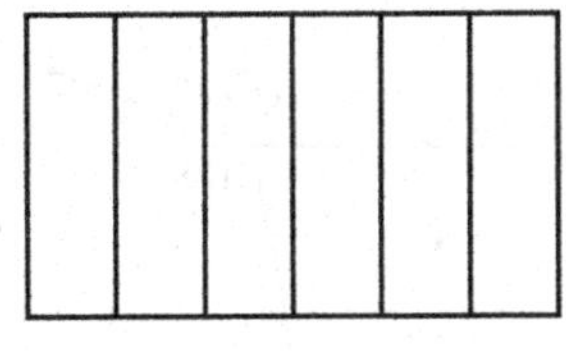

f 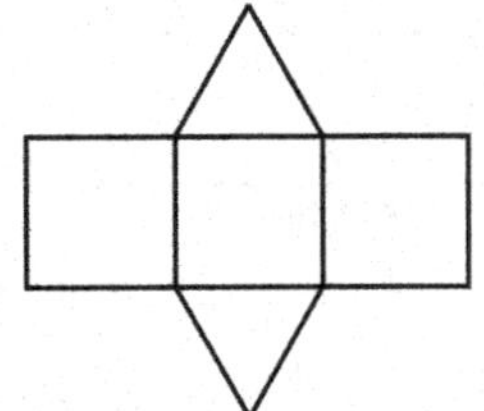

8.2.14 Use and read maps accurately

Understand scale on maps

Remember

The scale on maps is sometimes written as a ratio. It tells us the relationship between the distances on the map and the distances in real life.

For example: 1:100 000 means 1 cm on the map represents 1 km in real life, since 1 km = 100 000 cm.

1 Convert these measurements to the same unit of measurement and re-write the scale as a ratio.

a	1 cm = 2 km	**b**	2 cm = 5 km	**c**	1 cm = 2.5 km	**d**	1 cm = 500 m
e	1 cm = 750 m	**f**	2 cm = 1 km	**g**	2 cm = 100 m	**h**	5 cm = 20 km
i	5 cm = 250 m	**j**	2 cm = 50 m	**k**	1 cm = 40 m	**l**	1 cm = 1.5 km

2 The scale used on a map is 1:75 000. Calculate the real distances in kilometres for these distances measured from the map.

a	2 cm	**b**	5 cm	**c**	10 cm	**d**	0.5 cm	**e**	30 cm	**f**	6 cm
g	15 cm	**h**	12 cm	**i**	2.5 cm	**j**	4.5 cm	**k**	20 cm	**l**	22 cm

3 The scale used on a map is 1:5 000 000. Calculate the real distances in kilometres for these distances measured from the map.

a	5 cm	**b**	0.5 cm	**c**	8 cm	**d**	2 cm	**e**	10 cm	**f**	3.5 cm
g	7.5 cm	**h**	1.2 cm	**i**	12.5 cm	**j**	4.6 cm	**k**	11.5 cm	**l**	9.4 cm

4 The scale used on a map is 1:50 000. Calculate the real distances in kilometres for these distances measured from the map.

a	4 cm	**b**	10 cm	**c**	3 cm	**d**	5 cm	**e**	8 cm	**f**	15 cm
g	7 cm	**h**	11 cm	**i**	0.5 cm	**j**	2.5 cm	**k**	7.5 cm	**l**	4.5 cm

5 If the scale used on a map is 1:200 000, how many centimetres would represent these actual distances?

a	4 km	**b**	10 km	**c**	1 km	**d**	20 km	**e**	500 m	**f**	5 km
g	15 km	**h**	1500 m	**i**	2.5 km	**j**	7 km	**k**	12 km	**l**	9.5 km

6 If the scale used on a map is 1:8000, how many centimetres would represent these actual distances?

a	400 m	**b**	160 m	**c**	40 m	**d**	800 m	**e**	200 m	**f**	1 km
g	0.5 km	**h**	2.5 km	**i**	240 m	**j**	1.2 km	**k**	0.6 km	**l**	300 m

7 If the scale used on a map is 1:2 500 000, how many centimetres would represent these actual distances?

a	100 km	**b**	200 km	**c**	25 km	**d**	75 km	**e**	150 km	**f**	500 km
g	12.5 km	**h**	62.5 km	**i**	87.5 km	**j**	275 km	**k**	225 km	**l**	137.5 km

Assessment Space and Shape

1 Calculate the area of these shapes correct to two decimal places.

a

b
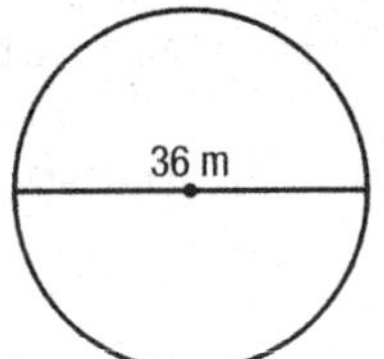

c
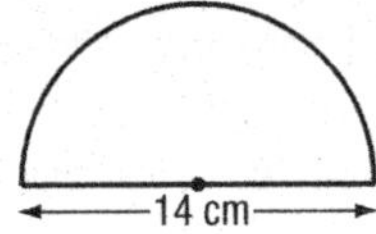

d
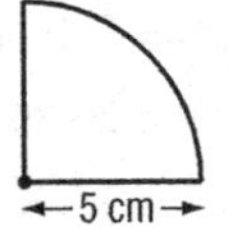

e
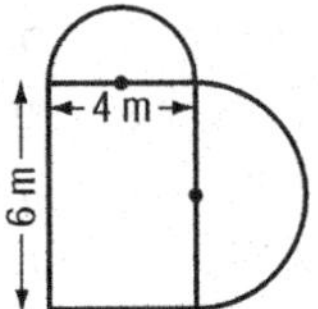

f
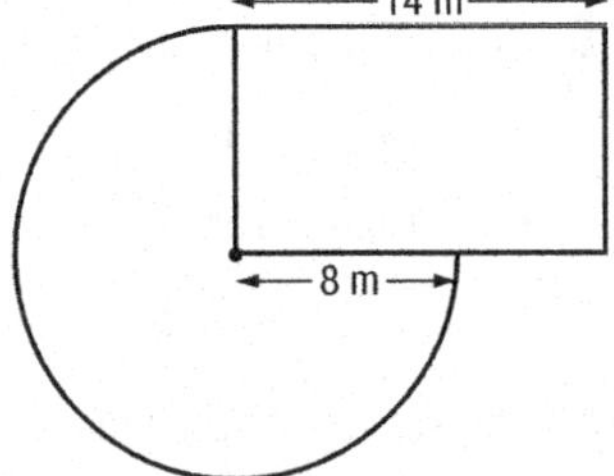

2 Calculate the shaded area of each diagram correct to two decimal places.

a
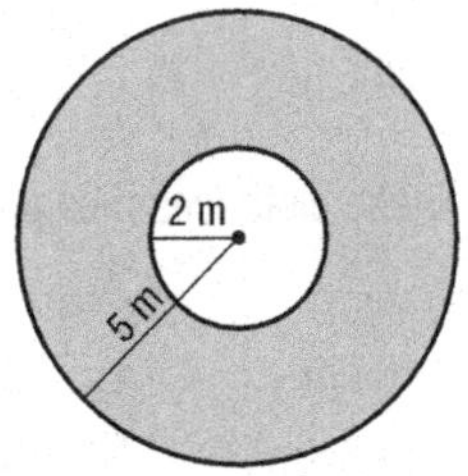

b
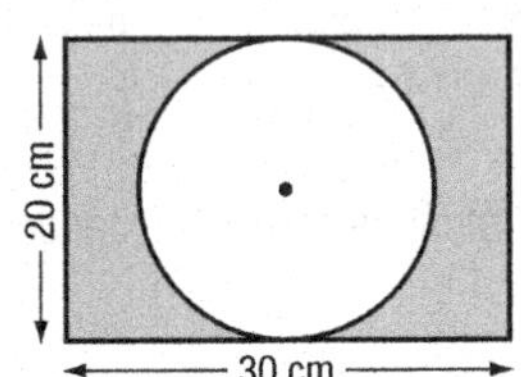

c
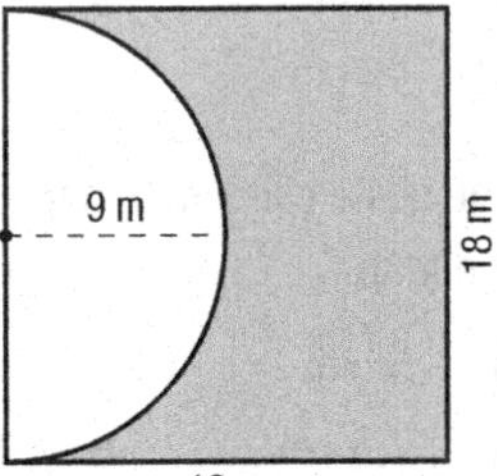

3 Calculate the volume of each three-dimensional shape.

a
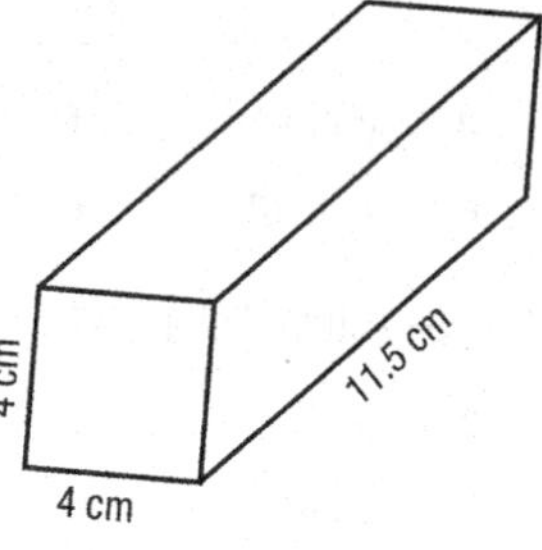

b
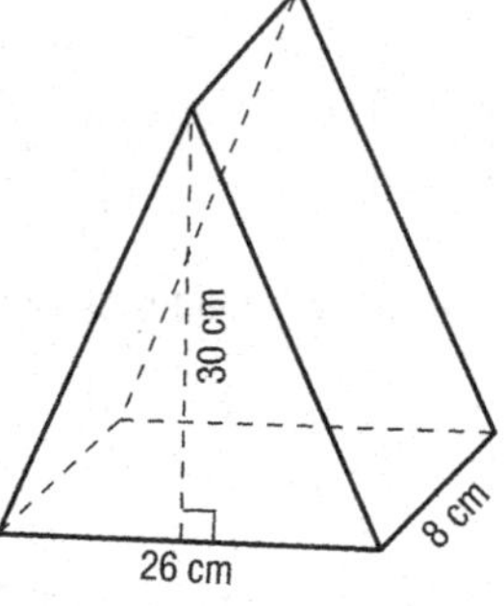

c
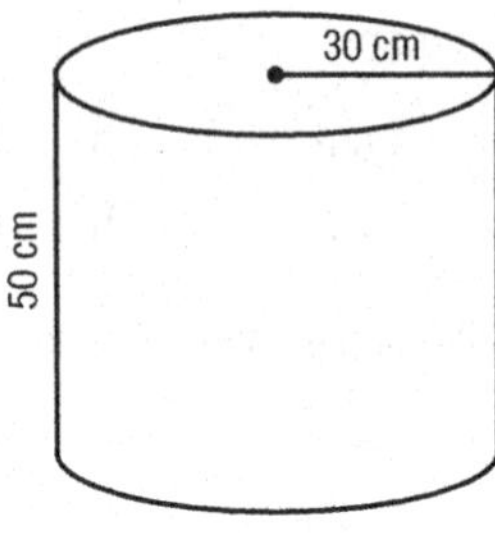

d
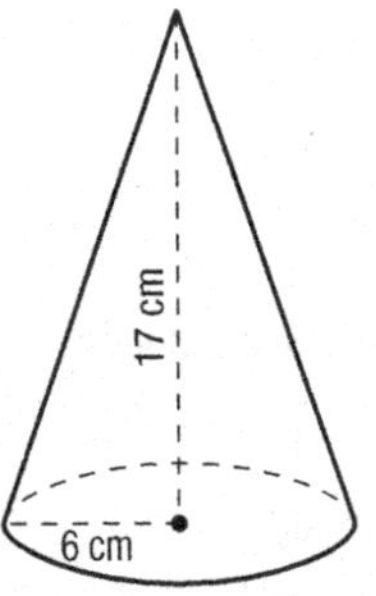

e
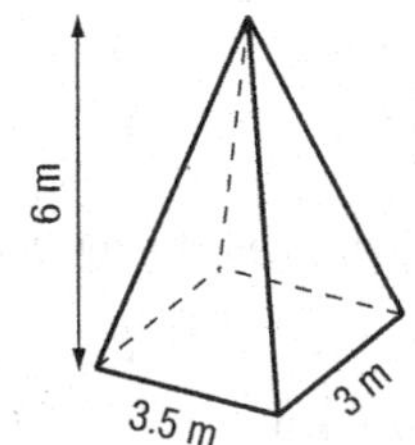

f
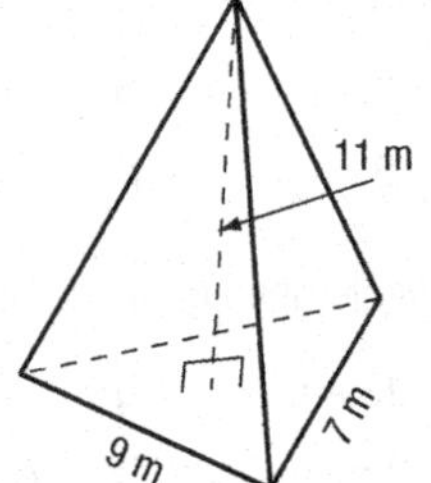

Assessment Space and Shape

4 What would be the volume of containers with these capacities?

a	746 mL	b	0.387 L	c	8.6 kL	d	6500 L
e	4.25 L	f	0.85 kL	g	455 mL	h	1765 mL

5 What would be the capacity of containers with these volumes?

a	218 m^3	b	9 m^3	c	64 cm^3	d	3485 cm^3
e	17.5 m^3	f	44 cm^3	g	31.2 m^3	h	0.4 m^3

6 Calculate the diameter of circles that have these radii.

a	6.5 cm	b	3.7 mm	c	11.4 mm	d	8.8 cm
e	12.6 cm	f	29.8 m	g	23.65 m	h	17.95 m

7 Calculate the radius of circles that have these diameters.

a	20.8 cm	b	13.9 cm	c	74.6 mm	d	115.5 mm
e	127.7 m	f	106.36 m	g	93.84 m	h	51.08 cm

8 Calculate the circumference of circles with these measurements to two decimal places.

a	radius of 9.7 cm	b	diameter of 12.8 cm	c	radius of 8.45 cm
d	diameter of 2.6 m	e	radius of 0.75 m	f	diameter of 1.36 m
g	radius of 15.38 cm	h	diameter of 26.45 m	i	radius of 19.17 cm

9 Calculate the diameter and the radius of circles with these circumferences to two decimal places.

a	88 cm	b	225 m	c	193 cm	d	137 m
e	206 cm	f	114 cm	g	258 cm	h	63 cm
i	159 m	j	233 m	k	188 m	l	104 m

10 The scale used on a map is 1:400 000. Calculate the real distances in kilometres for these distances measured from the map.

a	3 cm	b	10 cm	c	6.5 cm	d	0.75 cm	e	5 cm	f	8.5 cm

11 If the scale used on a map is 1:25 000, how many centimetres would represent these actual distances?

a	500 m	b	2 km	c	1.25 km	d	0.75 km	e	125 m	f	1.5 km

Assessment Space and Shape

12 Calculate the missing angle in each shape.

a

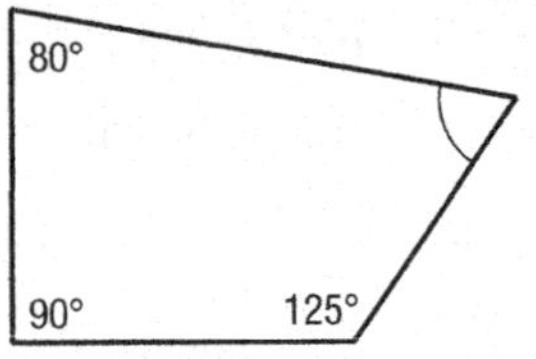

b

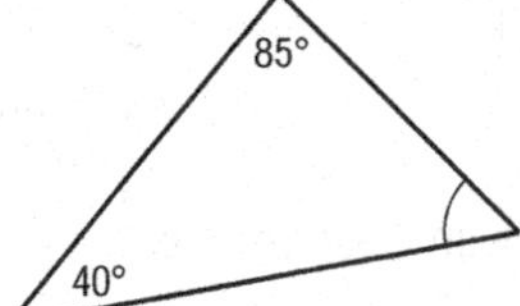

c

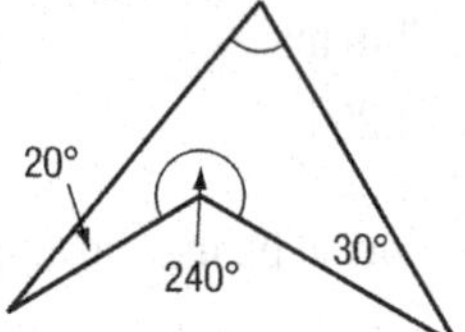

13 What is the sum of the internal angles of these shapes?

a a pentagon **b** an octagon **c** a decagon

14 Calculate the missing angle in each diagram.

a

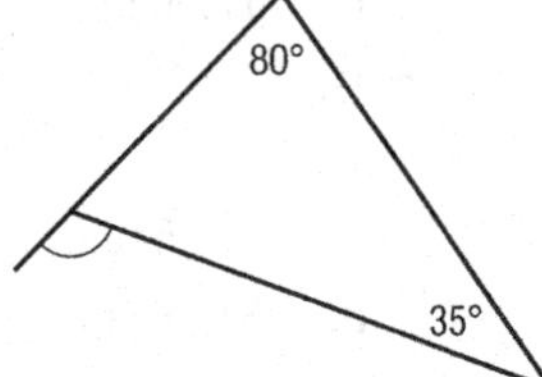

b

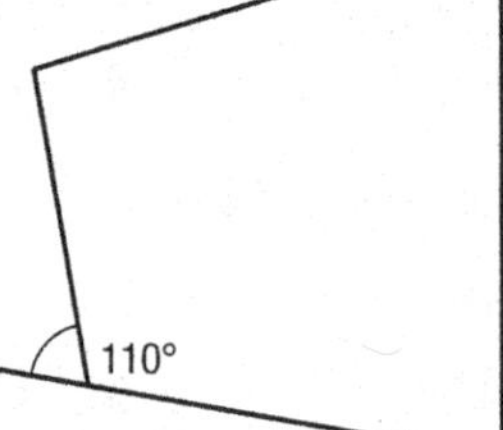

c

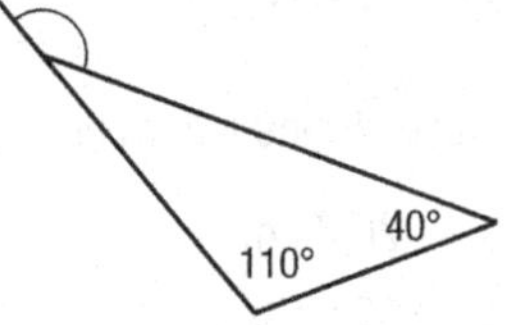

15 Determine the value of the pronumerals in each diagram.

a

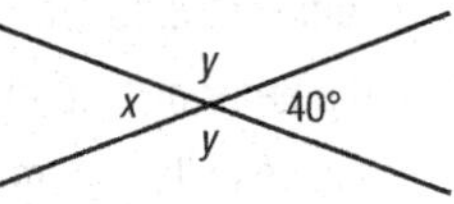

b

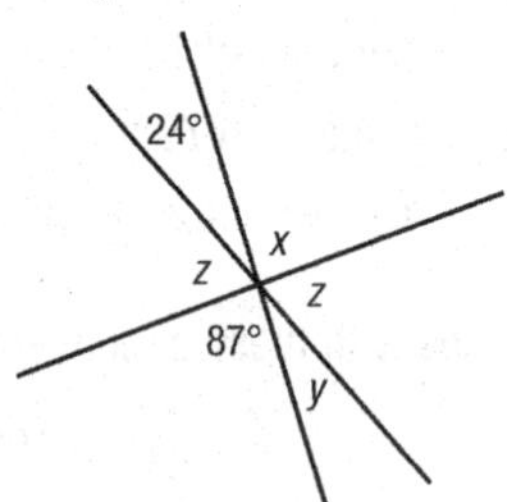

c

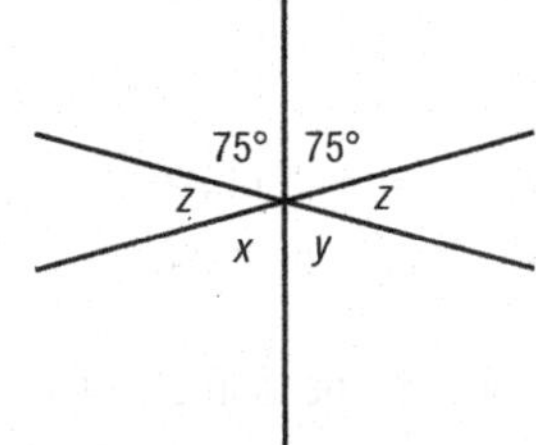

d

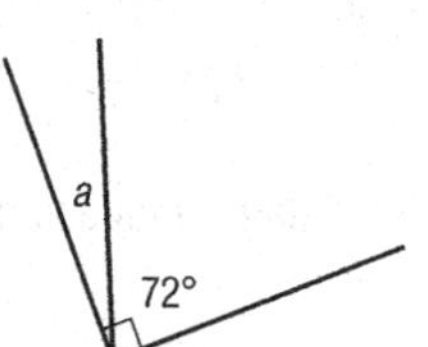

e

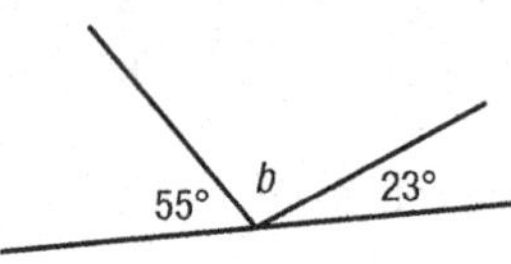

f

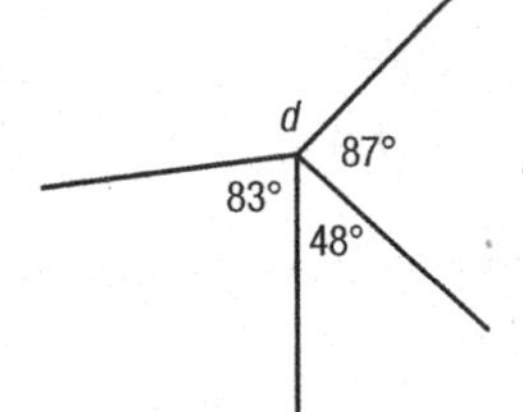

g

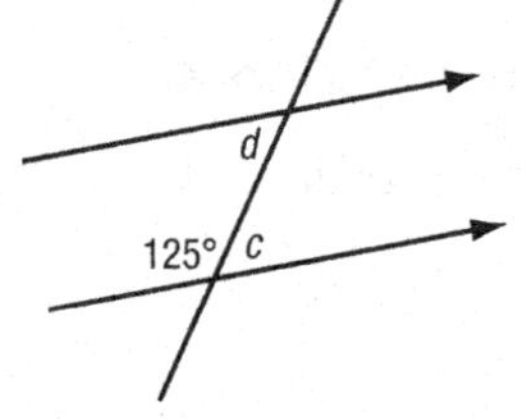

h

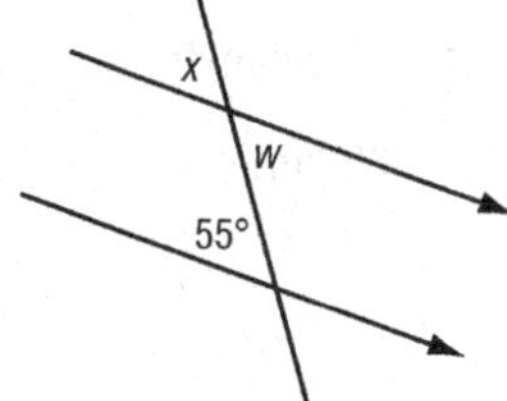

i

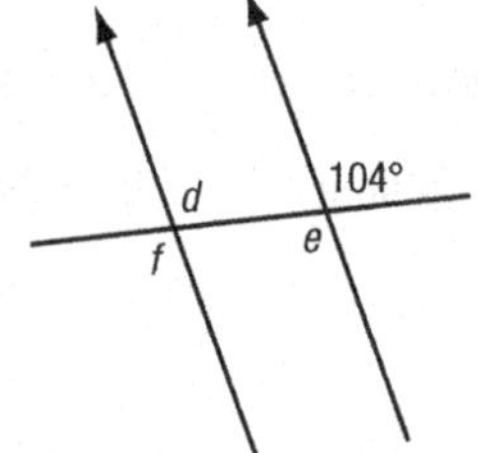

Strand Measurement

8.3.1 Solve real life weight problems with confidence and competence

Solve weight problems

1 Calculate the weight of an adolescent who weighed:

- **a** 35.7 kg at age 12 and gained 7.75 kg over the next two years
- **b** 48.5 kg at age 14 and gained 5.9 kg in the next year
- **c** 32 kg at age 13 and increased her weight by 20% over the next two years
- **d** 41 kg at age 14 and increased her weight by 15% in the next year
- **e** 37.6 kg at age 13 and gained 4.5 kg in the next year and then increased this by a further 10% the year after

2 Copy and complete this table, which shows the estimated weight gain of infants from birth to five months.

	Birth weight	Loss of 10% birth weight soon after birth	Increase birth weight by 30%	Increase birth weight by 50%	Double birth weight at 5 months
a	3100 g				
b	2.95 kg				
c	3.2 kg				
d	2750 g				

3 An airline company loads 247 bags onto a plane. The mean weight of each bag is 20.55 kg.

- **a** What is the total weight of the 247 bags?
- **b** How much less than 5500 kg is this?

4 The total weight of baggage checked in for 63 passengers is 1159.2 kg. What is the mean weight of each passenger's bag?

5 An airline charges K23.40 per kilogram for excess baggage. How much will it cost for these amounts of excess baggage?

a 7 kg **b** 2.5 kg **c** 4 kg **d** 9 kg **e** 5.5 kg **f** 12 kg

6 A passenger pays a total of K453.75 for excess baggage. If the passenger has 16.5 kg of excess baggage, how much is he paying per kilogram?

7 Regina has 27.15 kg of luggage. She is allowed no more than 22.5 kg. By how much does she have to reduce her luggage to avoid paying an excess baggage charge?

8 This chart shows information on the yield that can be expected per hectare (ha) of various crops in certain regions.

Crop	Tonnes/ha	Crop	Tonnes/ha
Carrots	3.5 tonnes	Taro	8 tonnes
Peanuts	4.3 tonnes	Bananas	10.5 tonnes
Cocoa	1.75 tonnes	Palm oil	16.7 tonnes

Calculate the expected yield if landowners have these crops planted.

- **a** 12 hectares of peanuts
- **b** 6.5 hectares of carrots
- **c** 4.5 hectares of bananas
- **d** 5 hectares of cocoa trees
- **e** 2.75 hectares of taro
- **f** 15 hectares of palm oil trees

9 Calculate the average yield per hectare for landowners who have these total yields.

- **a** 13 hectares of peanuts with a total yield of 56.5 tonnes
- **b** 9 hectares of bananas with a total yield of 42.75 tonnes
- **c** 1.5 hectares of taro with a total yield of 9.75 tonnes
- **d** 16 hectares of cocoa trees with a total yield of 27.6 tonnes

10 A landowner who grows peanuts and taro wants to increase his yield per hectare. He currently gets 3.8 tonnes of peanuts per hectare and 6.9 tonnes of taro per hectare. If he manages to increase his yield by 5%, what will his yield per hectare be for each crop?

11 A landowner who had 22 hectares planted with palm oil trees found that his yield per hectare decreased from 17.2 tonnes to 16.7 tonnes over a one-year period.

- **a** What was the landowner's total yield before the decrease?
- **b** What was his total yield after the decrease?
- **c** What percentage has the landowner's yield decreased by?

12 A delivery company loads packed boxes into a truck. Copy and complete this chart, which shows the number and weight of each type of packed box.

	Number of boxes	Weight per box	Total weight
a	8	17.45 kg	
b	6	23.8 kg	
c	11		172.37 kg
d	5		143 kg
e	16	9.37 kg	
f	9		354.6 kg
g	15		188.55 kg
h	7	17.58 kg	

8.3.4 Recognise relationships between location and time

Use time zones to determine time

Remember

Time is measured east or west from the prime meridian (longitude 0°). Time along the prime meridian is Greenwich Mean Time (GMT). All of Papua New Guinea is in the same standard time zone.

This chart shows the number of hours that cities around the world are ahead or behind Greenwich Mean Time.

Port Moresby	+10	Auckland	+12	Perth	+8
Los Angeles	−8	New York	−5	Sydney	+10
Tokyo	+9	Rome	+1	Johannesburg	+2
Mexico City	−6	Vancouver	−8	Jakarta	+7

1 Complete the gaps.

a When it is 6 a.m. in Port Moresby it is ______ in Auckland.
b When it is 1 p.m. in Sydney it is ______ in Perth.
c When it is 10 a.m. in Tokyo it is ______ in Rome.
d When it is 5 p.m. in Vancouver it is ______ in New York.
e When it is 2:30 p.m. in Jakarta it is ______ in Port Moresby.
f When it is 11:30 a.m. in Johannesburg it is ______ in Mexico City.
g When it is 9:30 p.m. in Los Angeles it is ______ in Sydney.
h When it is 3:15 p.m. in New York it is ______ in Vancouver.

2 When it is 1:15 p.m. on Thursday in Greenwich, calculate the day and standard time in the following places:

a Perth　b Mexico City　c Tokyo　d Rome
e Los Angeles　f Port Moresby　g New York　h Jakarta

3 When it is 8 p.m. on Sunday in Greenwich, calculate the day and standard time in the following places:

a Vancouver　b Sydney　c Auckland　d Johannesburg
e Port Moresby　f Perth　g Mexico City　h Rome

4 When it is 11:30 p.m. in Port Moresby on Monday, calculate the day and standard time in the following places:

a Tokyo　b Mexico City　c Jakarta　d Auckland
e Johannesburg　f Vancouver　g Los Angeles　h Sydney

5 When it is 6:30 a.m. in Port Moresby on Friday, calculate the day and standard time in the following places:

a Auckland　b Perth　c Rome　d New York
e Tokyo　f Sydney　g Vancouver　h Jakarta

Calculate with time

1 A time schedule showed that Jim had worked the following hours for the week.

Monday	8:20 a.m. – 3:45 p.m.
Tuesday	7:45 a.m. – 4:20 p.m.
Wednesday	8:05 a.m. – 3:50 p.m.
Thursday	8:15 a.m. – 4:35 p.m.
Friday	7:50 a.m. – 4:15 p.m.

- **a** Which day did Jim work the most hours?
- **b** How many hours did Jim work for the week?
- **c** If Jim was paid K15.50 per hour, what would he make for the week?

2 These are the arrival times for various flights. If each flight takes $1\frac{3}{4}$ hours, what time will they have to depart?

a 1120 **b** 1530 **c** 0825 **d** 2015 **e** 1350 **f** 2210
g 1655 **h** 2140 **i** 1045 **j** 1805 **k** 0935 **l** 1415

3 Copy and complete this film schedule, which shows the start, duration and finish times of the films at a cinema.

	Start time	Duration	Finish time
a	2:50 p.m.	89 minutes	
b		70 minutes	6:15 p.m.
c	10:10 a.m.	75 minutes	
d	6:20 p.m.		7:35 p.m.
e		108 minutes	11:50 p.m.
f	11:40 a.m.	115 minutes	

4 If each game goes for 1 hour and 35 minutes, how much longer will games that have been going for these times take to complete?

a 28 minutes **b** 16 minutes **c** 57 minutes **d** 43 minutes **e** 76 minutes
f 85 minutes **g** 39 minutes **h** 48 minutes **i** 21 minutes **j** 64 minutes

5 This fishing trip schedule shows the start and finish times of the various trips available through one company.

Trip	Start time	Finish time
A	0720	1345
B	0900	1215
C	1150	1645
D	1330	1740
E	1615	2010
F	1530	1945

- **a** Which trip takes the longest time?
- **b** Which trip takes the shortest time?
- **c** How long does Trip C take?
- **d** How long does Trip E take?
- **e** If you want a trip that takes between 4 and 5 hours, which ones could you choose?
- **f** If Trip F has a delayed start of 50 minutes, what will its new start and finish times be?

Solve time word problems

1 An athletics competition started at 1047 and finished 5 hours and 37 minutes later. What time did it finish?

2 A person spends 50 minutes at the gymnasium four days every week. How much time does the person spend at the gymnasium

a each week? **b** each month? **c** each year?

3 A tennis match finished at 2023. If the match lasted 2 hours and 52 minutes, what time did it start?

4 If you have five lessons a week of mathematics and each lesson is 70 minutes long, what is the total time spent in mathematics lessons in

a a week? **b** three weeks? **c** eight weeks?

5 It takes a teacher 4 hours and 35 minutes to mark 25 maths tests. What is the average time taken to mark each test?

6 A sales assistant earns K7.35 per hour. How much will the sales assistant earn if she works these hours?

a 5 hours **b** 7.5 hours **c** 18 hours **d** 15 hours

7 Sharon spent 85 minutes coaching a swimming squad on Monday, 2 hours 35 minutes on Tuesday, 115 minutes on Wednesday and $2\frac{3}{4}$ hours on Thursday.

a How much time did Sharon spend coaching swimming over the four days?

b How much more time did she spend coaching on Thursday than on Monday?

c How much less time did she spend coaching on Wednesday than on Tuesday?

8 It is recommended that beef be cooked for 60 minutes per kilogram, lamb for 50 minutes per kilogram and pork for 80 minutes per kilogram. Copy and complete this table of cooking times.

Kilograms

		1	1.5	2	2.5	3	3.5
a	**Beef**	60 minutes					
b	**Lamb**	50 minutes					
c	**Pork**	80 minutes					

9 In a 4 × 200-metre relay, the time run by the first runner was 36.5 seconds, the second runner ran 33.9 seconds, the third runner ran 35.6 seconds and the last runner ran 32.8 seconds.

a How long did it take to run the race?

b How much faster was the last runner than the first runner?

c What was the mean time for each 200-metre run?

Assessment Measurement

1 The total load allowed in a passenger elevator in an office building is 690 kg. Which of the following groups of passengers would be allowed to travel in the lift?

- **a** eight people with an average weight of 75 kilograms
- **b** ten people with an average weight of 72 kilograms
- **c** seven people with an average weight of 67.5 kilograms
- **d** nine people with an average weight of 70.5 kilograms
- **e** twelve people with an average weight of 63 kilograms

2 Calculate the average yield per hectare for landowners who have these total yields.

- **a** 15 hectares of bananas with a total yield of 139.5 tonnes
- **b** 7 hectares of peanuts with a total yield of 32.9 tonnes
- **c** 12 hectares of palm oil trees with a total yield of 189.6 tonnes
- **d** 13.5 hectares of taro with a total yield of 128.25 tonnes
- **e** 22 hectares of cocoa trees with a total yield of 40.7 tonnes

3 These are the arrival times for different flights. If each flight takes 85 minutes, what time will they have to depart in order to arrive on time?

a 1455 **b** 2025 **c** 0935 **d** 1040 **e** 1705 **f** 1150
g 1910 **h** 0830 **i** 2315 **j** 1245 **k** 1520 **l** 2140

4 What time will it be 1 hour and 40 minutes after these times?

a 8:45 a.m. **b** 4:20 p.m. **c** 11:10 a.m. **d** 2:17 p.m.
e 10:23 a.m. **f** 11:48 p.m. **g** 7:14 p.m. **h** 5:06 a.m.

5 What time will it be 2 hours and 25 minutes before these times?

a 9:26 a.m. **b** 8:47 p.m. **c** 3:41 p.m. **d** 8:09 a.m.
e 5:33 p.m. **f** 10:18 p.m. **g** 7:54 a.m. **h** 1:36 p.m.

6 Calculate the duration between these pairs of times.

a 10:44 a.m. and 1323 **b** 1138 and 4:19 p.m. **c** 6:36 p.m. and 2152
d 0916 and 12:45 p.m. **e** 7:11 a.m. and 1434 **f** 1751 and 11:38 p.m.
g 2:39 p.m. and 1806 **h** 1623 and 9:49 p.m. **i** 10:56 p.m. and 0224

7 Add these periods of time together.

- **a** 78 minutes + 2 hours 18 minutes + 53 minutes + 1 hour 37 minutes
- **b** 1 hour 22 minutes + 46 minutes + 1 hour 51 minutes + 95 minutes

Strand Chance and Data

8.4.1 Interpret information presented statistically

Find the mean, median, mode and range

Remember

The **mean** is found by adding a set of scores together and dividing by the number of scores.

The **median** is the middle score or number of a set of scores or numbers. It is found by putting the numbers in order. If there is an even number of data, the median is halfway between the two middle scores.

The **mode** is the score or number that occurs most often in a set of scores or numbers. There can be more than one mode in a set. If each score or number appears only once, then there is no mode.

The **range** is the spread between the highest and lowest scores and is found by subtracting the lowest score from the highest score.

1 Copy and complete this table. Where necessary, round answers to one decimal place.

	Data set	Mean	Median	Mode	Range
a	6, 9, 3, 8, 11, 19, 8				
b	10, 17, 26, 18, 10, 14				
c	7.6, 4.5, 5.1, 9.8, 6.5				
d	12.3, 9.5, 12.3, 7.4, 4.1				
e	25, 34, 19, 19, 25, 33, 25				
f	16, 11, 8, 5, 16, 13, 9				
g	44, 65, 79, 45, 36, 71				
h	7, 5, 8, 7, 8, 2, 9, 8, 8				
i	2.7, 4.1, 2.9, 3.6, 4.1, 1.5				
j	64, 87, 31, 64, 77, 38, 45				
k	79, 51, 55, 55, 63, 74, 18				
l	107, 96, 48, 119, 96, 32				
m	68, 43, 81, 95, 74, 43, 18				
n	1.3, 7.4, 7.4, 3.7, 3.7				
o	114, 99, 127, 94, 52, 127				

2 Find the mean of the following sets of numbers.

a all numbers between 30 and 40

b all numbers between 40 and 50

c the odd numbers between 30 and 50

d the even numbers between 30 and 50

Calculate relative frequency

Help Box

A relative frequency is the fraction of times an answer or result occurs. It is an approximation of the probability of an event occurring. To calculate relative frequency, divide the frequency by the sample size.

$$\text{Relative frequency} = \frac{\text{frequency}}{\text{sample size}}$$

Example: The results of a test showed that 7 out of 30 students got a score of 10, so the relative frequency can be calculated by dividing the frequency (7) by the sample size (30): $7 \div 30 = 0.23$, so 23% of the students who did the test got a score of 10.

Relative frequencies can be written as fractions, decimals or percentages.

1 Copy and complete these frequency tables by calculating the relative frequency of each value.

a Spelling test scores

Score	Frequency	Relative frequency
5	3	$3 \div 35 = 0.09$ (9%)
6	5	
7	8	
8	12	
9	5	
10	2	
Totals	35	

b Mathematics test scores

Score	Frequency	Relative frequency
1–20	2	
21–40	3	
41–60	15	
61–80	27	
81–100	25	
Totals		

c Weight (kg) of animals

Score	Frequency	Relative frequency
0–5 kg	6	
6–10 kg	10	
11–15 kg	5	
16–20 kg	18	
21–25 kg	9	
26–30 kg	8	
30+ kg	5	
Totals		

d Maximum temperatures for April

Score	Frequency	Relative frequency
31°	2	
30°	3	
29°	7	
28°	5	
27°	8	
26°	4	
25°	1	
Totals		

2 Make a frequency table, like the previous tables, for the following shoe sizes.

8	9	7	7	8	$8\frac{1}{2}$	$7\frac{1}{2}$	9	7	$7\frac{1}{2}$	8	$7\frac{1}{2}$	$8\frac{1}{2}$	7	$8\frac{1}{2}$	8
$6\frac{1}{2}$	$7\frac{1}{2}$	8	8	7	9	$8\frac{1}{2}$	$7\frac{1}{2}$	$7\frac{1}{2}$	$8\frac{1}{2}$	$7\frac{1}{2}$	$6\frac{1}{2}$	$7\frac{1}{2}$	$8\frac{1}{2}$	$7\frac{1}{2}$	8

a What percentage of the shoe sizes are $7\frac{1}{2}$?

b What percentage of the shoe sizes are 8?

Make predictions from sample data

Help Box

A random sample can be taken when it is not practical to survey an entire population. Predictions about the entire population are made from the results of the sample.

Example: A sample of 60 students out of a school population of 300 students was tested for a hearing problem, and 4 of the 60 students were found to have a hearing problem. To predict the number of the population who might have this hearing problem we use the formula: $\frac{\text{number with the characteristic in the sample}}{\text{sample size}} \times \text{population size.}$

$\frac{4}{60} \times 300 = \frac{1}{15} \times 300 = \frac{300}{15} = 20$

So, 20 students out of the school population of 300 (or 6.7%) are predicted to have a hearing problem.

1 From a school of 500 students, a random sample of 100 was selected to find out how they travelled to school. Of those selected, 84 walked to school.

- **a** What is the size of the population?
- **b** How many are in the sample?
- **c** What fraction of the sample walks to school?
- **d** Predict the number of students in the school population who walk to school.

2 Sixty people are randomly selected from a group of 200 and nine are found to be left-handed.

- **a** What is the size of the population?
- **b** How many are in the sample?
- **c** What fraction of the sample is left-handed?
- **d** Predict the number of people from the larger group who are left-handed.

3 From a group of 1200 tourists, a random sample of 250 was asked about the type of accommodation they preferred. Of those asked, 180 said they preferred to stay at hotels.

- **a** What is the size of the population?
- **b** What is the sample size?
- **c** What fraction of the sample preferred to stay at hotels?
- **d** Predict the number of tourists from the larger group who would prefer to stay at hotels.

4 Fifty people are randomly selected from the 275 who attended the screening of a new movie. Out of the 50 people, 43 said they liked the movie.

- **a** What is the size of the population?
- **b** What is the sample size?
- **c** What fraction of the sample liked the movie?
- **d** Predict the number of people out of those who attended who liked the movie.

Measure angles to interpret pie charts

1 This pie chart represents an amount of K1200 to be used to pay bills. Use a protractor to measure each angle, then work out the fraction and the value of each share.

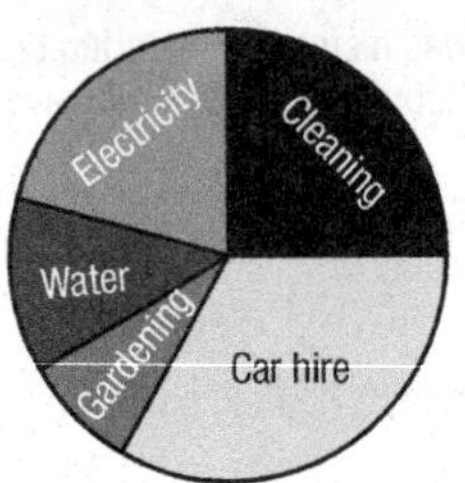

	Bill to be paid	Angle (degrees)	Fraction	Value of share
a	Electricity			
b	Water			
c	Car hire			
d	Cleaning			
e	Gardening			

2 This pie chart represents a survey of 60 students to choose an activity to do at a holiday camp. Measure each angle, then work out the fraction and the number of students who chose each activity.

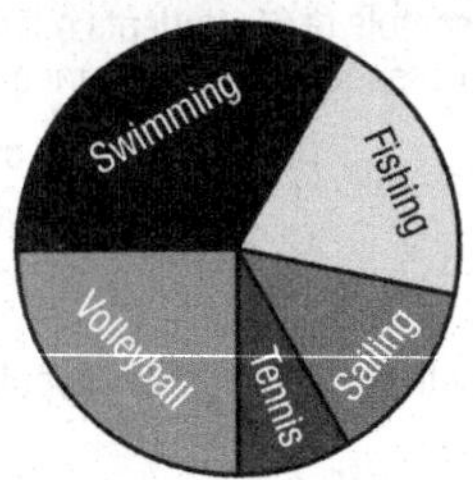

	Activity	Angle (degrees)	Fraction	Number of students
a	Swimming			
b	Tennis			
c	Volleyball			
d	Fishing			
e	Sailing			

3 This pie chart shows the country of birth of 240 people. Measure each angle, then work out the fraction and the number of people from each country.

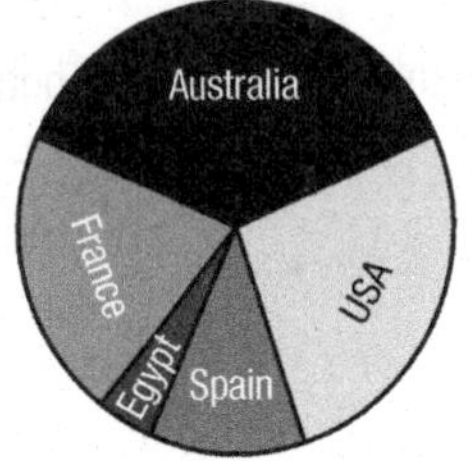

	Country of birth	Angle (degrees)	Fraction	Number of people
a	Australia			
b	USA			
c	France			
d	Spain			
e	Egypt			

4 This pie chart shows the results of a survey of the colour of 45 cars at a car park. Measure each angle, then work out the fraction and the number of cars of each colour.

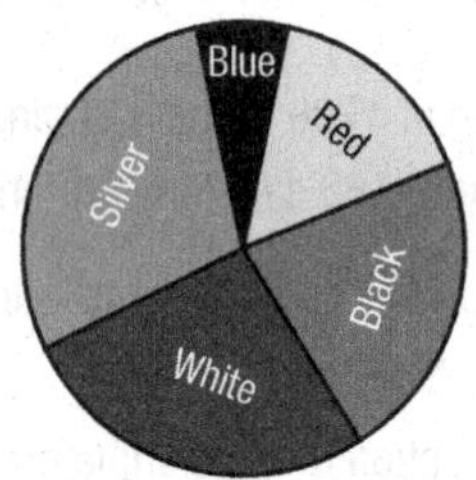

	Colour	Angle (degrees)	Fraction	Number of cars
a	White			
b	Silver			
c	Blue			
d	Black			
e	Red			

8.4.3 Explore the social implications of probability

Calculate experimental probability

Remember

Experimental probability is found by conducting an experiment. The relative frequency of an event occurring is the experimental probability of that event occurring. Experimental probability is calculated in the same way as relative frequency and is expressed as a fraction, decimal, ratio or percentage of the total number of trials.

$$\text{Relative frequency} = \frac{\text{frequency of an event}}{\text{total number of trials}}$$

1 This table shows the results of spinning a spinner with four different coloured sections. Calculate the experimental probability of each colour being spun, and write as fractions reduced to their lowest terms.

Colour	Red	Blue	Green	Yellow
Frequency	30	35	15	20

2 An 8-sided dice was rolled 200 times and the results are shown in this table.

Number	1	2	3	4	5	6	7	8
Frequency	23	26	19	27	21	29	24	31

a Calculate the experimental probability of each number being rolled. Write your answers as percentages correct to one decimal place.

b Calculate the relative frequency of rolling an even number.

c Calculate the relative frequency of rolling a number larger than 4.

3 Of 90 people in a village, 54 lived until at least 70 years of age.

a What is the relative frequency of someone from this village living until at least 70 years old? Write your answer as a percentage.

b What is the relative frequency of someone from this village not living until 70 years old?

4 This table shows the results of an experiment where a person chose one counter at a time from a bag containing red, yellow and blue counters. After each turn, the counter is returned to the bag. This process is repeated 60 times.

Colour	Red	Yellow	Blue
Frequency	17	22	21

Calculate the experimental probability of each colour being chosen. Write your answers as percentages correct to one decimal place.

5 This table shows the results of selecting one card at a time from a pack of white, red, black and green cards. After each turn, the card is returned to the pack and the pack is shuffled.

Colour	White	Red	Black	Green
Frequency	36	47	43	24

a Calculate the experimental probability of each colour being chosen as a percentage.

b What is the relative frequency of choosing either a white or a green card?

Calculate theoretical probability

Remember

Theoretical probability is found by noting all possible outcomes and determining how likely the given outcome is.

$$\text{Theoretical probability} = \frac{\text{number of successful outcomes}}{\text{total number of possible outcomes}}$$

1 What is the theoretical probability of choosing the following balls from a bag containing five red balls, three blue balls and seven green balls?

a a blue ball
b a red ball
c a green ball
d a red or blue ball
e a green or red ball
f a blue or green ball

2 A card is selected randomly from a box containing three orange cards, five purple cards and four yellow cards. Calculate the theoretical probability of selecting:

a an orange card
b a yellow card
c a purple card
d a blue card
e a purple, yellow or orange card

3 When rolling a ten-sided dice, calculate the theoretical probability of rolling:

a a number larger than 3
b an even number
c a prime number
d a number smaller than 5
e a number other than 8
f an 8, 9 or 10

4 Calculate the theoretical probability of the children in a two-child family being:

a two girls
b a boy and a girl
c two boys
d at least one boy
e at least one girl
f a girl and then a boy

5 Two hundred tickets numbered 1 to 200 were sold in a lucky number contest. Calculate the theoretical probability of these numbers being drawn:

a	any of the 200 numbers	**b**	a number above 50	**c**	a number ending in 0
d	a number starting with 2	**e**	a number with a 5 in it	**f**	a 3-digit number

6 A total of 1000 raffle tickets are sold. What is the probability of winning first prize if you have the following number of tickets?

a	five tickets	**b**	20 tickets	**c**	25 tickets
d	one ticket	**e**	40 tickets	**f**	100 tickets

7 A standard deck of 52 playing cards is shuffled and one card is selected from the top of the deck. Calculate the probability of the card being:

a	an ace	**b**	a diamond	**c**	the jack of spades
d	a 5 or a 6	**e**	a black 8	**f**	the 4 of hearts

8 Three different coins are tossed at the same time. Calculate the probability that the coins will land with these sides showing:

a	three heads	**b**	two heads and one tail	**c**	no heads
d	two tails and one head	**e**	at least one head	**f**	at least two tails

8.4.7 Estimate results of calculations

Use estimation methods

Remember

There are many situations where we do not need an exact answer, but only an approximate answer.

1 Round these large numbers to the nearest 100 000.

a	2 517 386	**b**	12 782 861	**c**	5 607 258	**d**	37 709 914
e	4 862 483	**f**	28 655 319	**g**	33 402 165	**h**	8 718 283
i	15 225 094	**j**	62 168 033	**k**	6 282 601	**l**	76 438 146

2 Use estimation to determine if the answer to each fraction addition is more or less than 1.

a	$\frac{2}{3}+\frac{5}{8}$	**b**	$\frac{3}{5}+\frac{1}{10}$	**c**	$\frac{2}{7}+\frac{3}{8}$	**d**	$\frac{4}{9}+\frac{2}{3}$	**e**	$\frac{5}{6}+\frac{1}{3}$	**f**	$\frac{3}{4}+\frac{1}{3}$
g	$\frac{2}{9}+\frac{2}{3}$	**h**	$\frac{1}{8}+\frac{2}{9}$	**i**	$\frac{2}{5}+\frac{1}{2}$	**j**	$\frac{7}{10}+\frac{3}{8}$	**k**	$\frac{5}{12}+\frac{1}{4}$	**l**	$\frac{1}{2}+\frac{1}{6}$

3 Use estimation to determine if the answer to each fraction addition is more or less than 2.

a	$\frac{4}{7}+1\frac{1}{4}$	**b**	$\frac{3}{4}+\frac{14}{15}$	**c**	$1\frac{1}{10}+\frac{2}{3}$	**d**	$\frac{4}{9}+\frac{2}{3}$	**e**	$\frac{1}{3}+1\frac{5}{6}$	**f**	$\frac{11}{12}+\frac{6}{7}$
g	$1\frac{2}{7}+\frac{1}{2}$	**h**	$\frac{10}{7}+\frac{12}{5}$	**i**	$\frac{3}{4}+1\frac{1}{3}$	**j**	$1\frac{3}{10}+\frac{4}{5}$	**k**	$\frac{5}{8}+\frac{7}{7}$	**l**	$\frac{1}{2}+1\frac{7}{12}$

4 Estimate answers by rounding to the first digit. For example: 5316 × 487 becomes 5000 × 500 = 2 500 000.

a	7233 × 225	**b**	4852 × 578	**c**	6787 × 713	**d**	5874 × 634
e	3695 × 289	**f**	8428 × 677	**g**	3234 ÷ 51	**h**	8864 ÷ 38
i	6173 ÷ 43	**j**	5716 ÷ 32	**k**	7449 ÷ 67	**l**	4852 ÷ 58

5 Calculate exact answers to Question 4. Use your estimated answers to check if your exact answer is reasonable.

6 Round these decimal numbers to the nearest hundredth (two decimal places).

a	51.638	**b**	23.855	**c**	76.075	**d**	33.562
e	92.081	**f**	47.367	**g**	19.166	**h**	67.914
i	38.277	**j**	29.863	**k**	56.154	**l**	85.528
m	16.075	**n**	64.687	**o**	31.283	**p**	70.664

7 Round these decimal numbers to the nearest tenth (one decimal place).

a	36.725	**b**	3.5183	**c**	72.861	**d**	48.364
e	8.3628	**f**	26.097	**g**	8.4289	**h**	61.055
i	54.427	**j**	6.7214	**k**	63.514	**l**	82.185
m	33.609	**n**	25.572	**o**	2.4355	**p**	40.786

Assessment Chance and Data

1 Find the mean and the median of these sets of scores.

a 16, 9, 23, 21, 17, 20, 27, 14, 22

b 65, 60, 54, 67, 85, 59, 72, 55

c 8.5, 3.7, 5.2, 8.1, 6.5, 4.9, 7.3, 8.2

d 9.7, 3.8, 5.9, 10.2, 8.8, 7.4, 11.6

2 Find the mode and the range of these sets of scores.

a 45, 38, 43, 41, 51, 38, 39, 40, 38

b 87, 96, 100, 71, 87, 105, 84, 79, 71

c 58, 63, 27, 66, 63, 59, 60, 62, 70

d 2.3, 4.6, 2.1, 2.1, 4.8, 8.7, 7.2, 5.9

3 From a group of 2000 people who attended a trade fair, 360 were randomly selected and asked if they currently smoked cigarettes. Of those selected, 38 said that they did smoke cigarettes.

a What is the population size?

b What is the sample size?

c What fraction of the sample smoked cigarettes?

d Predict the number of people from the larger group who smoke cigarettes.

4 Forty birds are randomly selected from an enclosure containing 200 birds and six of these are found to have a fungal infection.

a What is the population size?

b What is the sample size?

c What fraction of the sample has the fungal infection?

d Predict the number of birds in the enclosure that have the fungal infection.

5 A six-sided dice was rolled 180 times and the results are shown in this table.

Number	1	2	3	4	5	6
Frequency	29	32	27	29	33	30

a Calculate the experimental probability of each number being rolled. Write your answers as percentages correct to one decimal place.

b Calculate the relative frequency of rolling an odd number.

c Calculate the relative frequency of rolling a number smaller than 3.

6 A biscuit is randomly selected from a box containing nine chocolate biscuits, five ginger biscuits and four shortbread biscuits. Calculate the theoretical probability of selecting:

a a ginger biscuit

b a chocolate biscuit

c a shortbread biscuit

d a shortbread or chocolate biscuit

e a chocolate or ginger biscuit

7 A total of 500 raffle tickets are sold. What is the probability of winning first prize if you buy these numbers of tickets?

a one ticket

b eight tickets

c five tickets

d fifteen tickets

e two tickets

f twelve tickets

Strand Patterns and Algebra

8.5.2 Recognise and use patterns in processes

Use algebraic rules to complete tables

Use each formula to find the missing values.

1 $b = \frac{a}{3}$

a	30	12	21	45	18	36
b						

2 $m = 2c + 5$

c	8	13	11	20	17	26
m						

3 $y = 5x - 3$

x	12	9	40	22	16	51
y						

4 $d = h - 15$

h	30	43	21	54	27	62
d						

5 $s = 50 - t$

t	20	9	12	25	37	19
s						

6 $w = y^3$

y	10	5	12	20	3	11
w						

7 $f = 5(b - 4)$

b	16	11	24	8	12	7
f						

8 $a = 4(g + 15)$

g	24	31	17	28	14	35
a						

9 $m = \frac{4k}{5}$

k	10	5	15	20	40	25
m						

10 $p = 100 - 3v$

v	4	11	7	33	21	16
p						

11 $z = 9 + 2x$

x	26	17	22	31	47	19
z						

12 $q = r^2 + 8$

r	8	12	6	20	11	15
q						

13 $e = \frac{s}{4} - 2$

s	20	16	8	60	28	44
e						

14 $x = 45 - 3w$

w	9	6	12	15	7	11
x						

15 $a = \frac{d}{3} + 12$

d	21	60	27	6	54	39
a						

16 $h = 2(5f + 3)$

f	10	5	12	8	24	60
h						

Find algebraic rules

Find the rule that connects the two letters in each table of values and write it as an algebraic rule.

1

x	4	11	7	39	24	18
y	40	110	70	390	240	180

2

a	14	9	17	42	35	67
b	21	16	24	49	42	74

3

c	50	33	45	21	84	38
g	41	24	36	12	75	29

4

d	49	21	70	84	35	56
s	7	3	10	12	5	8

5

h	8	12	6	10	16	25
k	64	144	36	100	256	625

6

m	5	10	3	12	25	19
p	14	29	8	35	74	56

7

w	4	9	11	7	18	10
z	42	92	112	72	182	102

8

f	5	3	9	2	8	6
n	125	27	729	8	512	216

9

p	64	81	52	37	90	43
r	52	69	40	25	78	31

10

c	15	11	20	9	30	17
d	90	66	120	54	180	102

11

x	32	24	100	72	88	64
y	8	6	25	18	22	16

12

g	6	10	5	11	9	20
h	37	101	26	122	82	401

13

k	64	81	36	144	100	49
m	8	9	6	12	10	7

14

s	8	10	5	7	25	16
t	38	48	23	33	123	78

15

n	62	91	55	47	83	70
p	47	76	40	32	68	55

16

y	4	9	10	6	12	3
z	14	79	98	34	142	7

17

j	9	15	21	11	8	30
k	81	135	189	99	72	270

18

b	80	32	72	48	120	136
d	10	4	9	6	15	17

Find algebraic rules and use them to complete tables

Find the rule that connects the two letters in each table of values. Write it as an algebraic rule and then use it to complete the tables.

1

e	20	9	14	57	36	78
f	24	13	18			

2

b	31	49	65	22	83	54
c	20	38	54			

3

s	4	11	8	10	30	12
a	16	121	64			

4

t	20	44	32	84	96	52
v	5	11	8			

5

x	6	11	9	10	30	8
y	72	132	108			

6

w	8	12	20	11	30	17
d	21	29	45			

7

g	15	32	50	42	27	65
h	29	63	99			

8

k	16	47	80	39	54	62
m	36	67	100			

9

r	72	93	61	88	76	107
p	22	43	11			

10

a	150	30	370	210	90	120
b	15	3	37			

11

n	8	64	1000	125	27	216
z	2	4	10			

12

f	3	10	8	15	7	5
g	60	200	160			

13

c	7	10	30	4	15	80
d	31	40	100			

14

p	9	3	10	6	12	40
k	43	13	48			

15

w	89	56	71	92	43	35
x	81	48	63			

16

f	5	10	8	11	30	9
m	26	101	65			

17

y	7	11	4	15	9	6
z	48	120	15			

18

h	54	29	37	14	83	62
b	79	54	62			

8.5.3 Manipulate simple algebraic expressions and solve real life problems

Write and interpret multiplication and division in algebraic expressions

Remember

In algebra, the multiplication and division signs are not generally used. Instead of writing, for example, $5 \times a$, we write $5a$. The number is always written first. Instead of writing, for example, $b \div 3$, we write $\frac{b}{3}$.

1 Re-write these expressions without multiplication or division signs.

a	$4 \times a \times b$	**b**	$d \times 7$	**c**	$x \times y \times z$	**d**	$12 \times p$
e	$g \div 8$	**f**	$23 \div w$	**g**	$ab \div 5$	**h**	$17 \div xy$
i	$6 \times z \div 3 \times w$	**j**	$2 \times a \times b \div c$	**k**	$c \times 9 \times g$	**l**	$m \div 4$
m	$v \times 8 \div 3$	**n**	$7 \times x + 5 \times y$	**o**	$2d \times 3$	**p**	$s \div 5 - m \div 3$
q	$4 \times g \times 2 \times h$	**r**	$m \times n \times 6 \times 3$	**s**	$5 \times 6ab$	**t**	$d \div 5 - 3 \times g$
u	$8 \times m \times n \div s$	**v**	$3 \times g \times 5 \div 2h$	**w**	$y \div 4 + 3 \times a$	**x**	$k \times 6 - 12 \div z$

Remember

When a pronumeral is multiplied by itself, we write it in index notation; for example, $a \times a$ is written as a^2. Similarly, $b \times b \times b$ is written as b^3.

2 Write the following algebraic expressions using index notation.

a	$4 \times m \times m$	**b**	$9 \times d \times d \times d$	**c**	$a \times a \times 5$
d	$2 \times 5 \times y \times y$	**e**	$2a \times 2a$	**f**	$5 \times y \times z \times y \times z \times y$
g	$4 \times h \times 3 \times h \times h$	**h**	$b \times b \times b \times b \times 8$	**i**	$3d \times 2d \times d$
j	$5 \times a \times a \times a \times b$	**k**	$4c \times 5c \times 2c$	**l**	$w \times w \times w \times w \times 7$
m	$g \times h \times 3 \times h \times g \times h$	**n**	$5y \times y \times 2d \times d \times d$	**o**	$7k \times k \times 3k$
p	$6 \times f \times g \times h \times h \times 2$	**q**	$8m \times 3m$	**r**	$2w \times 2w \times 3w$

Remember

When multiplying like terms, we add the powers; for example, $b^3 \times b^5 = (b \times b \times b) \times (b \times b \times b \times b \times b) = b^8$. Similarly, $5b^2 \times 2b^3 = 10b^5$.

When dividing like terms, we subtract the powers; for example, $\frac{y^5}{y^2} = \frac{(y \times y \times y \times y \times y)}{(y \times y)} = y^3$. Similarly, $\frac{6y^4}{2y^2} = 3y^2$.

A term without an index number means the term to the power of 1; for example, $\frac{6d^3}{2d}$ is the same as $\frac{6d^3}{2d^1}$, which is $3d^2$; $3a^4 \times 2a$ is the same as $3a^4 \times 2a^1$, which is $6a^5$.

3 Multiply.

a $y^5 \times y^3$
b $c^6 \times c^2$
c $k^4 \times k^3$
d $b^3 \times b^2$
e $2g^3 \times 3g^2$
f $4k^3 \times 5k^3$
g $3a \times 6a^2$
h $5d^4 \times 5d$
i $m \times 2m^5$
j $2w^2 \times 2w$
k $4h^5 \times 7h$
l $3p^2 \times 3p^2$
m $a^3 \times 2a^2 \times 2a^3$
n $3b \times 3b \times 2b^2$
o $2y^2 \times y^2 \times 3y^2$
p $k^5 \times 2k \times 2k^4$

4 Divide.

a $\frac{a^7}{a^4}$
b $\frac{d^6}{d^2}$
c $\frac{w^5}{w^3}$
d $\frac{y^8}{y^3}$
e $\frac{9b^4}{3b^2}$
f $\frac{8k^3}{2k^2}$
g $\frac{10h^6}{2h^4}$
h $\frac{12d^6}{4d^2}$
i $\frac{8w^4}{4w}$
j $\frac{15g^6}{5g}$
k $\frac{8x^4}{2x^2}$
l $\frac{16b^7}{4b^3}$
m $\frac{12y^9}{3y}$
n $\frac{6d^8}{6d^2}$
o $\frac{18w^3}{6w}$
p $\frac{4k^2}{4k}$

Simplify by adding or subtracting like terms

Help Box

In algebra, letters are often used as a substitute for numbers. These letters are called pronumerals. Pronumerals can be added and subtracted if they have like terms.

Example 1: $2y$ and $7y$ are like terms so $2y + 7y = 9y$ and $7y - 2y = 5y$

Example 2: $4ab$ and $3ba$ are like terms so $4ab + 3ba = 7ab$ and $4ab - 3ba = ab$.

The terms $5a$ and $2a^2$, $6x$ and $3y$, $2d$ and 5 are not like terms so cannot be added or subtracted.

1 Which of these pairs are like terms? Write yes or no for each one.

a $2b$ and b^2
b $4xy$ and $3yx$
c $2cd$ and cd
d $4gh$ and $4gh^2$
e $3mn$ and $3m^2n$
f $6abc$ and $5abc$
g 9 and $6d$
h $4k$ and $8kw$
i $7v^2$ and v^2
j $3xyz$ and $8zyx$
k $6b$ and $-7b$
l $5a^3$ and $2a^2$
m $-2ds$ and $5sd$
n $16f$ and $33f$
o $17y$ and $-8y$
p $4mp$ and mp^2

2 Which of these algebraic expressions can be simplified by adding or subtracting like terms? Write yes or no for each one.

a $7a - 5$
b $2x + 6x$
c $3cd - cd$
d $8ab + 5ba$
e $4x^2 + 5x$
f $8z - 5s$
g $12dg + 7g$
h $15 - 6b$
i $2xy + 9xy$
j $11h - 6$
k $5mn + 6nm$
l $8y^2 - 3y^2$
m $10c - 7d$
n $9w + 4w$
o $-3d + 7d$
p $4a + 7a - 2a$

3 Simplify by adding or subtracting the like terms.

a	$m + m$	b	$d + d + d + d$	c	$ab + ab + ba$	d	$h + h + h$
e	$7b - 2b$	f	$8x - 4x - x$	g	$y - y$	h	$8d - d$
i	$9k + k + 3k$	j	$6gh - 3gh$	k	$5xw + 2wx$	l	$10g - 5g - 2g$
m	$12y + y$	n	$3x^2 + 4x^2$	o	$8p - 6p$	p	$7a^3 - a^3$
q	$v + v + v$	r	$8mn - 5nm$	s	$3s + 9s - 2s$	t	$7x - 4x + x$
u	$7d + d - 3d$	v	$2a^2 + 5a^2 - a^2$	w	$11gh - 7hg$	x	$12k - 2k - 3k$

4 Simplify these algebraic expressions by collecting like terms.
For example: $3d - 4y - d + 6y = 3d - d - 4y + 6y = 2d + 2y$

a	$3a + 4a - b$	b	$5w - 2w + 3x$	c	$6x + 5x + 7y$
d	$7t - t + 3c$	e	$10 - 4g - g$	f	$5h + 3 + 3h$
g	$6a + 5b + 2a$	h	$3m - m + 4n$	i	$6y + y - 5x$
j	$4k - 3 + 2k$	k	$3 + 5w + w$	l	$8d + 6 + 5d$
m	$7h + 3s - 4h$	n	$5a - 3b - a + 8b$	o	$6p + 5pq - 3p$
p	$4ab + 3a - ba + 2a$	q	$7xy + 3d + 4xy + 5d$	r	$5gh - 2g + 3gh - g$
s	$3cd + 2c - cd + 5c$	t	$8 + 3y - 4 + 3xy - y$	u	$3ab + 5 + 4xy - xy$
v	$5d + 3gh + 4d + 7$	w	$b + b^2 + 2b - 3b^2$	x	$5xy^2 - 2xy - xy^2 - 3xy$

Use the distributive law to expand algebraic expressions

Help Box

The distributive law is useful in algebra where the terms inside brackets are not always like terms. We use the distributive law to remove the brackets by multiplying each term inside the brackets by the term outside the brackets. This process is called expansion.

Example 1: $5(a + b) = 5 \times a + 5 \times b = 5a + 5b$

Example 2: $8(a - b) = 8 \times a - 8 \times b = 8a - 8b$

Remember to always put the number first in each expression; for example, $7h$ not $h7$.

1 Expand these addition expressions.

a	$7(x + y)$	b	$2(c + d)$	c	$5(w + z)$	d	$10(s + t)$
e	$3(g + h)$	f	$8(p + r)$	g	$11(n + m)$	h	$6(k + v)$
i	$5(y + 3)$	j	$3(x + 7)$	k	$9(a + 4)$	l	$4(b + 5)$
m	$d(c + 2)$	n	$a(b + 5)$	o	$g(h + 7)$	p	$w(y + 4)$
q	$m(4 + d)$	r	$5(7 + p)$	s	$y(5 + z)$	t	$9(5 + k)$
u	$(a + b)6$	v	$(w + g)4$	w	$(h + 3)8$	x	$(5 + y)3$

2 Expand these subtraction expressions.

a	$4(a-b)$	**b**	$7(f-g)$	**c**	$12(x-y)$	**d**	$3(m-n)$
e	$6(p-r)$	**f**	$9(c-d)$	**g**	$2(w-z)$	**h**	$5(s-t)$
i	$8(g-5)$	**j**	$4(h-6)$	**k**	$10(d-3)$	**l**	$6(f-8)$
m	$b(y-2)$	**n**	$p(r-7)$	**o**	$x(y-12)$	**p**	$v(w-3)$
q	$c(5-s)$	**r**	$3(4-g)$	**s**	$c(9-d)$	**t**	$8(2-m)$
u	$(g-h)7$	**v**	$(n-s)3$	**w**	$(a-8)2$	**x**	$(6-b)5$

3 Expand these expressions.

a	$5(4a+2)$	**b**	$7(3b-5)$	**c**	$2(6g+7)$	**d**	$8(5y-3)$
e	$3(6-4a)$	**f**	$9(2+5b)$	**g**	$4(10-4d)$	**h**	$6(3+2y)$
i	$4(2x+3z)$	**j**	$5(3c-5d)$	**k**	$7(6g+3h)$	**l**	$3(4m-5n)$
m	$3(5p-r)$	**n**	$7(9a+b)$	**o**	$11(6m-n)$	**p**	$9(4y+z)$
q	$6(5g+4h)$	**r**	$3(7f-2w)$	**s**	$8(3y+6d)$	**t**	$10(7c-5a)$
u	$(3k-7v)2$	**v**	$(4m+5p)4$	**w**	$(6t-2s)5$	**x**	$(8b+9d)3$

Remember

Letters are usually written in alphabetical order within a term.

Example 1: $x(5y+7z)=5xy+7xz$

Example 2: $2b(5a-3c)=10ab-6bc$

4 Expand.

a	$k(4a+5b)$	**b**	$a(3g-2h)$	**c**	$x(7y+4z)$	**d**	$d(5w-2n)$
e	$(3h-2g)m$	**f**	$(7x+4y)d$	**g**	$(5c-4d)g$	**h**	$(8a+3b)z$
i	$4w(3d+4b)$	**j**	$2k(3a-5b)$	**k**	$5y(4h+3g)$	**l**	$7d(5m-2k)$
m	$5x(2y-3w)$	**n**	$(4d+7c)2f$	**o**	$3n(8m-6p)$	**p**	$(2z+5y)4x$
q	$6t(3c+10d)$	**r**	$4g(5h-k)$	**s**	$3b(d+7k)$	**t**	$5n(6v-m)$

Write algebraic expressions

Remember

Algebra provides a shortcut method of writing a set of instructions for doing calculations; for example, the product of a and $b = ab$; the sum of g and h, multiplied by $3 = 3(g + h)$.

1 Write these in algebraic form.

- **a** the sum of x and y
- **b** the product of g and h
- **c** the difference between a and b
- **d** w divided by 7
- **e** the number 2 more than d
- **f** the number 5 less than z
- **g** twice k, add 7
- **h** 3 times m, subtract 2
- **i** twice the sum of s and t
- **j** half the product of y and z
- **k** 5 times g, subtract 3
- **l** 9 times n, add 6
- **m** 4 less than the product of x and y
- **n** 7 more than the quotient of c and d
- **o** the difference between p and q
- **p** k divided by 11, plus 4
- **q** multiply f by 6
- **r** divide m by 10
- **s** the number 8 less than g
- **t** the number 12 more than b

2 Write an algebraic expression for these instructions. Use brackets where necessary.

- **a** multiply the sum of b and d by 7
- **b** divide the sum of x and y by 5
- **c** divide w by g and then add 2
- **d** add p to q and then subtract 1
- **e** add a to 5 and then multiply by 8
- **f** subtract 3 from g, then multiply by 4
- **g** multiply m and n, then add 8
- **h** divide b by 4, then add 2
- **i** subtract h from 6, then divide by k
- **j** add 7 to d and then multiply by t
- **k** multiply the sum of 3 and y by w
- **l** divide the product of c and d by 2
- **m** add g to h, then divide by 5
- **n** divide the sum of 4 and k by y.
- **o** subtract d from 9, then multiply by 2
- **p** multiply b and 5, then add c

3 Write instructions for these algebraic expressions.

a	$z - 7$	**b**	$g + 8$	**c**	$4a$	**d**	$\frac{b}{3}$
e	xy	**f**	$p + q$	**g**	$k - 10$	**h**	$5g - 2$
i	$8 + 4d$	**j**	$6(g + h)$	**k**	$3(y - z)$	**l**	abc
m	$cd + 11$	**n**	$\frac{mn}{2}$	**o**	$s + \frac{t}{5}$	**p**	$\frac{g}{h} + 7$
q	$15 - 6d$	**r**	$k(c + 9)$	**s**	$4(v - 5)$	**t**	$f - \frac{n}{3}$

4 Write these instructions as algebraic expressions; for example, the first one is $a = 6b - 7$.

To find a:

a subtract 7 from 6 times b

b multiply d, f, and g together

c add 10 to c, then divide by 3

d divide x by y, then add 8

e multiply k by 4, then add d

f subtract 6 from p, then multiply by b

g add y to z, then multiply by 5

h add s, t and u together, then halve

i subtract 4 times d from 30

j divide g by 7, then subtract h

Expand and simplify algebraic expressions

Help Box

You have seen how the distributive law is used to remove brackets.

Example 1: $7(a + 3) + 5$ becomes $7a + 21 + 5$

This can then be simplified by collecting like terms, so $7a + 21 + 5$ becomes $7a + 26$.

Example 2: $a(3b - b + 4)$ becomes $3ab - ab + 4a$

This can then be simplified by collecting like terms, so $3ab - ab + 4a$ becomes $2ab + 4a$.

Expand and simplify each group of expressions by collecting like terms.

1

a $5(g + 5) - 7$

b $8(b - 1) + 12$

c $3(x + 4) + 15$

d $4(5 + 6p) - 2p$

e $7(3 + 5d) - 4d$

f $2(5y - 2) + 7y$

g $w(4b - 2b + 7)$

h $a(5d + d + g)$

i $y(w - 6 + 5w)$

j $4(m + 2) - 3m$

k $9(k + 6) - 4k$

l $7(2d - 3) + 5d$

m $10 + 3(g + 4)$

n $12 + 5(h - 2)$

o $4k + 8(k + 2)$

p $5p + 6(p - 3)$

q $8d + 3(d - 4)$

r $6s + 2(s + 5)$

s $6 + 3(f + 4)$

t $7(2a + 5a + a)$

u $4(g + 6g - 2g)$

v $3(xy + 4xy + y)$

w $2(5k + 3pq - pq)$

x $5(cd + 3cd - 2cd)$

2

a $6(5g + 9) - 5$

b $8(5k - 3) + 7$

c $4(11 + 6h) - 9$

d $4(y + 3) - 2y + 7$

e $7(x + 10) + 5x - 6$

f $3(3 + 4d) - 2d + 5$

g $6(9 - 4n) + 24n$

h $8(4 - 6g) + 50g$

i $11(3 - 5b) + 12b$

j $5(2 + 3p) + 2(4p - 1)$

k $4(3 + 4k) + 3(2k + 6)$

l $7(3a - 2) + 4(5 - 4a)$

m $4(3 - 2h) + 2(h + 5)$

n $8(2 - 3d) + 4(2d + 2)$

o $6(5 - 5w) + 5(6w + 3)$

p $y(4x - x + 5)$

q $m(7n + n - 8)$

r $3k(2d + 5d + 3)$

s $8 + 4(b + 3) - b + 2$

t $11 + 3(2y + 1) - 2y + 5$

u $5z + 20 + 6(3z - 2)$

v $3c(5 - 2d) + 5c$

w $4a(7 - 3k) - 8a$

x $2h(6 + 5g) + 10h$

Remember

Where a pronumeral outside a bracket is the same as a pronumeral within a bracket, then we use index notation to write the product; for example, $h(h + 7) = h^2 + 7h$; $3d(2d - 2) = 6d^2 - 6d$.

3 Expand and simplify these expressions, using index notation where necessary.

a	$x(x + 10x)$	**b**	$y(3y - y)$	**c**	$b(2b + 7b - 3b)$
d	$p(4p - p)$	**e**	$n(n + 3n + 2n)$	**f**	$2a(a + 5a - 2a)$
g	$5w(x + 5x + 3w)$	**h**	$2d(4d + e + 2e)$	**i**	$4a(5b + 2a - 3b)$
j	$n(8 + n) + 9n$	**k**	$3h(4 - 5h) + 7h$	**l**	$5g(2g + 4g)$
m	$7f(4f + 3f + 6)$	**n**	$3k + 2k(3k - k) + k$	**o**	$m(m + 3m) - 5m$
p	$5h + 2h(4h - h)$	**q**	$5w(w - a) + 2w^2$	**r**	$4p(2s - p + 2p) + 8$
s	$7y + g(g + 3g) - 2y$	**t**	$3c + 2c(2c - c) + c$	**u**	$8 + a(a + 5a - 3a)$
v	$n(n + 2n) + f(3f - f)$	**w**	$z(3z - z) + 2t(t + 4t)$	**x**	$3b(3b - b) + 4p(p + p)$

Use substitution to solve algebraic expressions

Help Box

We use substitution to find the value of an algebraic expression by replacing the letters with numbers. This process is called **evaluating** the algebraic expression.

Example: Evaluate the expression $2a + b$, if $a = 5$ and $b = 7$.
$2a + b = 2 \times 5 + 7 = 10 + 7 = 17$

1 If $w = 5$ and $y = 3$, evaluate these algebraic expressions.

a	$w + y$	**b**	$w - y$	**c**	wy	**d**	$\frac{y}{w}$
e	$3w + 4y$	**f**	$7w - 8y$	**g**	$3wy$	**h**	$5(w + y)$
i	$w(w + y)$	**j**	$6(2y + 4w)$	**k**	$w^2 - y^2$	**l**	$2wy + 6y$
m	$2y + 3wy$	**n**	$20 - wy$	**o**	$5(4w - 3y)$	**p**	$\frac{6w}{y}$

2 If $a = 2$ and $b = 7$, evaluate these algebraic expressions.

a	$4a + 5b$	**b**	$3ab$	**c**	$\frac{2b}{a}$	**d**	$\frac{a + 2b}{4}$
e	$b^2 - a^2$	**f**	$4(9a - 2b)$	**g**	$a(a + b)$	**h**	$2ab - b$
i	$40 - ab$	**j**	$\frac{7a}{b}$	**k**	$b^2 - 5a$	**l**	$\frac{10b}{a}$
m	$8b - 6a$	**n**	$5ab + 11a + 2b$	**o**	$3ab - 10a$	**p**	$3(5b - 2a)$

3 If $m = 8$ and $d = 5$, evaluate these algebraic expressions.

a	$3m - 2d + 6$	**b**	$2(m + 5) - 3(d + 2)$	**c**	$2dm + 3d + 5m$	**d**	$50 - dm$
e	$5(10m - 5d)$	**f**	$\frac{5m + 10}{2d}$	**g**	$6d - 2m - 4$	**h**	$3dm - d^2$
i	$m^2 + d^3$	**j**	$d^3 - m^2$	**k**	$2dm - 20$	**l**	$100 - dm$
m	$4(2m - 2d)$	**n**	$\frac{3dm}{10}$	**o**	$3m + 2dm$	**p**	$d(d + m)$

4 If $c = 4$, $g = 3$ and $h = 6$, evaluate these expressions.

a	$2cgh$	**b**	$6c + 4g + 7h$	**c**	$2gh + 3c$	**d**	$10h - cg$
e	$7(c + g + h)$	**f**	$5h + 5g - 6c$	**g**	$cg^2 - h^2$	**h**	$\frac{cgh}{3}$
i	$4(h + c) - 3g$	**j**	$5c + 2(2g - h)$	**k**	$10h - 4c - 5g$	**l**	$\frac{5cg}{h}$
m	$c^2 + g^2 + h^2$	**n**	$h(c + g)$	**o**	$2c(h - g)$	**p**	$11h + 5g - 10c$

5 If $k = 10$, $n = 4$ and $p = 2$, evaluate these expressions.

a	$\frac{4k}{n}$	**b**	$5p + kn$	**c**	$2kn - 10p$	**d**	$k^2 - p^4$
e	$k^2 + n^2 + p^2$	**f**	$8p + 4n + 3k$	**g**	$5(k + p - n)$	**h**	$2np - k$
i	$3k + 3(2n - p)$	**j**	$75 - 3np$	**k**	$3p(k - n)$	**l**	$p(k + n)$
m	$n^2 + kp$	**n**	$knp - 15p$	**o**	$\frac{5n}{k}$	**p**	$3(3n + k + 5p)$

6 If $v = 9$, $x = 8$ and $y = 7$, evaluate these expressions.

a	$5v - 3y$	**b**	$xy + vx$	**c**	$vy - 2x$	**d**	$4(v + x + y)$
e	$9v - 8x - y$	**f**	$v^2 + y^2$	**g**	$vxy - xy$	**h**	$2v + 5x - 3y$
i	$\frac{vy}{7}$	**j**	$v(x + y)$	**k**	$80 - vx$	**l**	$8y + 2x + 4v$
m	$x(x + y)$	**n**	$95 - y^2$	**o**	$2xy - 4v$	**p**	$v^2 + x^2 - y^2$

7 Copy and complete these tables by substituting into the given rule.

a $y = 4x - 5$

x	y
3	
8	
2	
10	
15	

b $m = n^2 + 3$

n	m
5	
11	
8	
12	
7	

c $a = \frac{b}{3}$

b	a
21	
36	
9	
60	
27	

d $g = 5h + 7$

h	g
4	
9	
3	
11	
20	

e $w = 45 - z$

z	w
6	
10	
4	
19	
15	

f $d = 36 - 4c$

c	d
7	
5	
2	
9	
6	

g $p = 9k - 5$

k	p
3	
8	
5	
10	
6	

h $t = s + 30$

s	t
20	
16	
14	
27	
13	

i $f = \frac{5e}{2}$

e	f
4	
10	
6	
12	
16	

j $c = 2b + 15$

b	c
7	
30	
8	
25	
17	

k $r = v^2 - 8$

v	r
5	
9	
3	
7	
10	

l $y = 5(w - 2)$

w	y
12	
15	
9	
22	
18	

m $q = 7(p - 5)$

p	q
17	
10	
8	
12	
6	

n $z = \frac{y^2}{2}$

y	z
4	
3	
10	
12	
5	

o $d = 6 + 4a$

a	d
8	
15	
9	
30	
22	

Identify factors and factorise algebraic expressions

Help Box

To find the highest common factor (HCF) of numbers, we need to list all the factors of each number, and then find the largest number that appears in all lists.

Example 1: Find the HCF of 15 and 24
List the factors of 15: 1, 3, 5 and 15
List the factors of 24: 1, 2, 3, 4, 6, 8, 12 and 24
The HCF (largest number in all lists) is 3, so the HCF of 15 and 24 is 3.

Example 2: Find the HCF of 20, 35 and 40
List the factors of 20: 1, 2, 4, 5, 10 and 20
List the factors of 35: 1, 5, 7 and 35
List the factors of 40: 1, 2, 4, 5, 8, 10, 20 and 40
The HCF of 20, 35 and 40 is 5.

1 Find the highest common factor of the following pairs of numbers.

a	8 and 40	**b**	18 and 30	**c**	36 and 16	**d**	50 and 15
e	28 and 56	**f**	24 and 64	**g**	15 and 42	**h**	12 and 33
i	45 and 30	**j**	72 and 48	**k**	81 and 54	**l**	66 and 121

2 Find the highest common factor of the following sets of numbers.

a	15, 24 and 30	**b**	8, 44 and 28	**c**	35, 98 and 77	**d**	36, 96 and 54
e	108, 45 and 72	**f**	132, 60 and 84	**g**	48, 80 and 16	**h**	56, 96 and 120

3 Use factors to help you solve these multiplications; for example, $52 \times 16 = 52 \times 8 \times 2 = 416 \times 2 = 832$.

a	47×15	**b**	85×24	**c**	63×18	**d**	78×25
e	94×32	**f**	67×21	**g**	141×45	**h**	184×36

Help Box

Algebraic expressions can be broken down into **factors**. Factorising is the opposite of expanding.

Example: Factorise the expression $6c + 12$

Step 1: Break the terms into their highest common factors: $6c + 12 = 2 \times 3 \times c + 2 \times 2 \times 3$

Step 2: Put the common factors of each term outside the brackets: $= 2 \times 3(c + 2)$

Step 3: Remove any multiplication signs: $= 6(c + 2)$

Step 4: Expand the answer to check its accuracy: $= 6c + 12$

4 Factorise these expressions.

a	$4h + 16$	**b**	$7w - 28$	**c**	$3b + 21$	**d**	$6y - 18$
e	$18a + 6$	**f**	$32d - 8$	**g**	$30k + 12$	**h**	$9m - 6$
i	$15 + 40z$	**j**	$56 - 24f$	**k**	$36 + 9y$	**l**	$66 - 44s$

5 Factorise these expressions.

a	$8x + 8y$	**b**	$3f - 3g$	**c**	$12a + 8b$	**d**	$28m - 21n$
e	$15w + 35k$	**f**	$26d - 14c$	**g**	$16s + 24t$	**h**	$36p - 28r$
i	$20mp + 70mp$	**j**	$64gh - 40gh$	**k**	$18bt + 4bt$	**l**	$42wy - 18wy$
m	$45pq + 25p$	**n**	$32ad - 8a$	**o**	$14v + 8vw$	**p**	$48s - 32ks$

Assessment — Patterns and Algebra

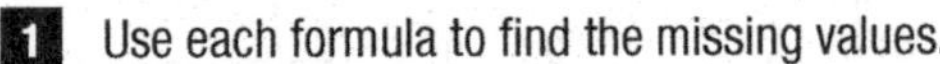

1 Use each formula to find the missing values.

a $y = 7z$

z	15	20	12	50	11	63
y						

b $h = 4g - 9$

g	7	15	25	12	30	16
h						

2 Find the rule that connects the two letters in each table of values. Write it as an algebraic rule and then use it to complete the tables.

a

m	21	15	45	60	18	54
n	7	5	15			

b

k	28	16	10	25	44	14
p	57	33	21			

3 Re-write these expressions without using multiplication or division signs.

a $6 \times c \times d$ **b** $w \times 7$ **c** $e \times f \times g$ **d** $y \times 18$

e $b \div 10$ **f** $12 \div x$ **g** $cd \div 7$ **h** $24 \div mn$

i $3t \times 4$ **j** $z \times 5 \div 2$ **k** $8 \times 2xy$ **l** $3 \times f \times 3 \times g$

m $5 \times p \times p$ **n** $c \times c \times 5$ **o** $3k \times 2k$ **p** $a \times a \times a \times a$

4 Simplify these algebraic expressions by collecting the like terms.

a $4y + 5y + 2w$ **b** $7d - 5d + 2d$ **c** $8m - 3m + 4n$ **d** $10s - s + 2t$

e $2 + 7g + g$ **f** $15 - 6w - w$ **g** $9v + 3vw - 5v$ **h** $a + a^2 + 4a$

5 Expand these expressions.

a $4(c + d)$ **b** $7(h + 6)$ **c** $3(a - b)$ **d** $k(p - 2)$

e $3(6g + 4)$ **f** $2(4d - 5)$ **g** $4(5m - 2n)$ **h** $(6f + 4g)3$

6 Expand and simplify these expressions.

a $3(h + 6) - 5$ **b** $5(2 + 3x) - 4x$ **c** $4(y + 2) + 10$ **d** $f(6g + g + h)$

e $w(z - 4 + 3z)$ **f** $d(4d - d)$ **g** $2k(3k + 2k + 7)$ **h** $3a(8a - 4a)$

7 If $a = 9$ and $b = 6$, evaluate these algebraic expressions.

a $2a + 3b$ **b** $4ab$ **c** $a^2 - b^2$ **d** $4(2a + b)$

e $3b + 2ab$ **f** $8a - 4b - 10$ **g** $100 - ab$ **h** $\frac{12a}{b}$

8 Factorise these expressions.

a $5g + 30$ **b** $9t - 18$ **c** $4w + 24$ **d** $28s - 7$

e $25 + 35x$ **f** $36 - 30y$ **g** $12a + 12b$ **h** $7c - 7d$

Important Facts

Length and perimeter

10 millimetres (mm) = 1 centimetre (cm)

100 centimetres (cm) = 1 metre (m)

1000 metres (m) = 1 kilometre (km)

1 mm = $\frac{1}{10}$ cm = 0.1 cm

1 cm = $\frac{1}{100}$ m = 0.01 m

1 m = $\frac{1}{1000}$ km = 0.001 km

Time

60 seconds = 1 minute

60 minutes = 1 hour

24 hours = 1 day

7 days = 1 week

2 weeks = 1 fortnight

52 weeks = 1 year

12 months = 1 year

365 days = 1 year

366 days = 1 leap year

10 years = 1 decade

100 years = 1 century

1000 years = 1 millennium

a.m. = ante meridiem – the period between 12 midnight and 12 noon

p.m. = post meridiem – the period between 12 noon and 12 midnight

12-hour time divides a day into two 12-hour periods: a.m. and p.m.

24-hour time divides a day into twenty-four 1-hour periods, starting at midnight

Capacity

1000 millilitres (mL) = 1 litre (L)

1000 litres = 1 kilolitre (kL)

1 mL = $\frac{1}{1000}$ L = 0.001 L

1 L = $\frac{1}{1000}$ kL = 0.001 kL

Weight

1000 grams (g) = 1 kilogram (kg)

1000 kilograms (kg) = 1 tonne (t)

1 g = $\frac{1}{1000}$ kg = 0.001 kg

1 kg = $\frac{1}{1000}$ t = 0.001 t

Money

100 toea (t) = 1 kina (K)

1t = $\frac{1}{100}$ K = K0.01

Area

10 000 square centimetres (cm^2) = 1 square metre (m^2)

10 000 square metres (m^2) = 1 hectare (ha)

100 hectares (ha) = 1 square kilometre (km^2)

Area of a rectangle = Length × Width

Area of a triangle = (Height × Base) ÷ 2

Area of a parallelogram = Base × Height

Area of a circle = πr^2 (r is the radius and π represents 3.14)

Temperature

Temperature is usually measured in degrees Celsius (°C). On the Celsius scale, water freezes at 0 degrees and boils at 100 degrees.

Volume

100 000 cubic centimetres (cm^3) = 1 cubic metre (m^3)

Volume of a rectangular prism = Length × Width × Height

Volume of a triangular prism = (Area of base) × Height

Volume of a cylinder = $\pi r^2 h$ (r = radius, h = height)

Volume of a cone = $\frac{\pi r^2 h}{3}$

Volume of a pyramid = $\frac{Ah}{3}$ (A = area of base, h = height)

Volume and capacity

1 mL = 1 cm^3; 1 L = 1000 cm^3;

10 L = 10 000 cm^3; 1000 L = 1 m^3

Circumference of circles

Circumference can be calculated by either of these formulae: $C = \pi D$ or $C = 2\pi r$ where C is circumference, D is diameter, r is radius and π represents 3.14

Directed numbers

Directed numbers are positive and negative whole numbers including zero.

A minus sign is always used to show negative numbers, for example –5, –10 and –127.

Positive numbers can be shown with or without a plus sign, for example +2 or 2, +17 or 17 and +36 or 36.

Index numbers

Positive index numbers (indices) are used to show that a number is multiplied by itself a certain number of times. For example, in 4^3 the index 3 shows that 4 is multiplied by itself 3 times: $4 \times 4 \times 4$.

An index number of zero always gives a result of 1. For example, $7^0 = 1$.

Negative index numbers are used to show how many times to divide by the number. For example, in 2^{-3} the index –3 shows that 1 is divided by 2 three successive times: $1 \div 2 \div 2 \div 2$.

Angles

Angles are measured in degrees.

A right angle = 90°

A straight angle = 180°

A full turn = 360°

Acute angles measure between 0° and 90°.

Obtuse angles measure between 90° and 180°.

Reflex angles measure between 180° and 360°.

The interior angles of a triangle total 180°.

The exterior angles of a triangle total 360°.

The interior angles of a quadrilateral total 360°.

The sum of the interior angles of a polygon can be found by the formula $(n - 2) \times 180°$, where n is the number of sides on the polygon.

Square roots and cube roots

The opposite of a square number is a square root. The symbol $\sqrt{}$ means square root. For example, $\sqrt{36} = 6$ because 6×6 (6^2) = 36.

The opposite of a cube number is a cube root. The symbol $\sqrt[3]{}$ means cube root.

For example, $\sqrt[3]{27} = 3$ because $3 \times 3 \times 3$ (3^3) = 27.

Order of operations

Do **brackets** first, then **of**, then **multiplication** and **division** in the order they appear from left to right, and finally **addition** and **subtraction** in the order they appear from left to right.

Answers

Strand: Number and Application

Page 2

1 97 881, 118 338, 127 265, 160 349, 376 929, 446 862, 506 146, 594 585, 604 553, 683 157, 780 273, **1 108 728**

2 289 676, 352 015, 408 943, 416 447, 512 820, 545 684, 548 775, 576 274, 620 631, 703 736, 708 559, **777 508**

3 169 760, 205 691, 292 737, 504 571, 513 539, 574 074, 616 382, 632 021, 1 075 843, 1 100 220, 1 185 136, **1 189 534**

4 361 172, 454 620, 678 652, 726 266, 737 959, 782 080, 850 997, 1 005 765, 1 045 631, 1 055 510, 1 108 116, **1 202 941**

Page 3

1 a 26 182, 24 658, 44 701, 261 304, 600
b 157 657, 106 817, 176 176, 129 829, 132 075
c 119 149, 68 309, 137 668, 168 337, 93 567
d 834 105, 783 265, 852 624, 546 619, 808 523
e 8928, 41 912, 27 447, 278 558, 16 654
f 171, 50 669, 18 690, 287 315, 25 411
g 249 788, 198 948, 268 307, 37 698, 224 206
h 89 917, 39 077, 108 436, 197 569, 64 335

2 a 31 084 b 75 716 c 117 806 d 59 431
e 291 151 f 157 145 g 207 276 h 137 252
i 287 616 j 132 414 k 327 461 l 262 919

Page 4

1 a 227 624 b 382 632 c 413 172 d 142 508
e 383 236 f 1 785 568 g 926 670 h 2 346 105
i 1 382 554 j 943 668 k 997 360 l 1 653 596
m 1 236 816 n 356 058 o 2 311 932 p 1 525 769
q 4 572 336 r 3 175 524 s 663 900 t 1 857 102
u 1 015 152 v 1 722 006 w 1 999 782 x 3 047 616

2 a 4 429 828 b 15 453 513 c 85 116

Page 5

1 a 8554 b 7945 c 6877 d 8476
e 17 396 f 28 163 g 42 783 h 63 218
i 2504 j 1694 k 4135 l 3476
m 6587 n 2858 o 4096 p 5445
q 1779 r 3817 s 2566 t 7284

2 a $950\frac{5}{8}$ b $1623\frac{2}{5}$ c $852\frac{2}{7}$ d $1542\frac{1}{3}$
e 6272 f $3856\frac{8}{9}$ g $1959\frac{5}{8}$ h $6819\frac{3}{5}$
i $73\,239\frac{1}{7}$ j 275 976 k 46 231 l $28\,830\frac{3}{4}$
m $2762\frac{5}{6}$ n $2313\frac{11}{25}$ o $1659\frac{7}{16}$ p $2045\frac{9}{16}$
q $1227\frac{6}{17}$ r $3271\frac{1}{15}$ s $2577\frac{17}{28}$ t $5015\frac{17}{19}$

Page 6

1 20 843

2 a 43 401 km b 101 269 km c 57 868 km d 130 203 km
e 86 802 km f 115 736 km

3 a K50 573 b K10 114.60

4 a 23 456 steps b 422 208 steps

5 a K20 808 b K13 872 c K41 616 d K31 212
e K62 424 f K52 020

6 a 7649 kg b 43 217 kg

7 a 614 819 b 1 058 299 c 212 433 d 1 248 999
e 1 061 338

8 K12 125.75

9 K6737

Pages 7–8

1 45, 4, 42, 32, 21, 40, 40, 12, 18, 72, 99, 144, 36, 0, 24, 54, 25, 88, 9, 0, 5, 22, 16, 64, 30, 36, 48, 16, 49, 40

2 7, 6, 9, 8, 1, 5, 8, 4, 7, 3, 9, 3, 6, 4, 5, 10, 11, 2, 12, 12, 8, 9, 6, 5, 5, 11, 12, 9, 3, 5

3 9, 72, 28, 120, 72, 54, 15, 24, 49, 18, 18, 100, 12, 36, 8, 0, 121, 24, 24, 45, 96, 16, 16, 56, 18, 54, 28, 48, 0, 30

4 10, 8, 5, 9, 4, 8, 10, 2, 9, 2, 1, 8, 3, 9, 10, 12, 9, 9, 3, 3, 3, 9, 5, 7, 4, 5, 9, 9, 11, 12

5 21, 72, 72, 35, 54, 56, 8, 0, 30, 32, 63, 24, 48, 55, 56, 25, 28, 27, 35, 12, 66, 64, 32, 0, 60, 36, 72, 70, 33, 42

6 54, 25, 18, 24, 56, 72, 84, 36, 36, 108, 20, 42, 45, 66, 96, 18, 21, 60, 7, 32, 132, 24, 24, 49, 35, 54, 88, 28, 72, 48

7 6, 5, 11, 6, 7, 10, 8, 1, 5, 3, 5, 5, 8, 9, 2, 9, 3, 6, 8, 7, 6, 4, 4, 10, 7, 8, 8, 9, 9, 3

8 54, 9, 16, 66, 0, 2, 24, 110, 36, 36, 35, 24, 49, 99, 72, 20, 9, 12, 120, 16, 18, 32, 64, 42, 35, 27, 81, 10, 45, 144

9 8, 3, 4, 10, 4, 8, 10, 12, 8, 6, 3, 5, 9, 9, 7, 11, 8, 12, 12, 11, 12, 9, 8, 10, 9, 9, 9, 2, 12, 12

10 32, 27, 2, 12, 30, 56, 18, 49, 72, 70, 11, 81, 24, 45, 20, 36, 56, 40, 99, 16, 48, 55, 18, 36, 24, 132, 0, 30, 35, 56

Page 9

1 a $<$ b $>$ c $<$ d $>$ e $>$ f $>$
g $>$ h $<$ i $>$ j $>$ k $<$ l $<$
m $<$ n $<$ o $<$ p $>$

2 a $1\frac{3}{4}$ b $1\frac{2}{3}$ c $3\frac{1}{5}$ d $1\frac{3}{4}$ e $3\frac{2}{5}$ f $2\frac{5}{8}$
g $2\frac{4}{7}$ h $\frac{14}{5}$ i $5\frac{1}{4}$ j $\frac{11}{7}$ k $4\frac{9}{10}$ l $1\frac{1}{5}$
m $3\frac{5}{7}$ n $\frac{13}{3}$ o $2\frac{4}{9}$ p $3\frac{7}{8}$ q $\frac{20}{6}$ r $\frac{23}{4}$

3 a–r Answers will vary. Teacher to check.

4 a $\frac{1}{10}, \frac{1}{4}, \frac{3}{5}, \frac{2}{3}, \frac{5}{3}$ b $\frac{1}{3}, \frac{3}{8}, \frac{7}{12}, \frac{7}{4}, 2\frac{3}{4}$
c $\frac{5}{8}, \frac{3}{4}, \frac{5}{6}, 1\frac{7}{8}, \frac{10}{5}$ d $\frac{2}{3}, \frac{5}{7}, \frac{4}{5}, \frac{9}{9}, 1\frac{2}{7}$
e $\frac{1}{8}, \frac{3}{7}, \frac{1}{2}, \frac{8}{9}, 1\frac{1}{3}$ f $\frac{4}{9}, \frac{7}{11}, \frac{9}{10}, 1\frac{1}{8}, \frac{8}{7}$
g $\frac{4}{7}, \frac{7}{10}, \frac{11}{12}, \frac{10}{7}, \frac{6}{3}$ h $\frac{4}{3}, \frac{13}{8}, 1\frac{3}{4}, 2\frac{1}{2}, \frac{15}{5}$
i $\frac{2}{9}, \frac{2}{7}, \frac{1}{2}, \frac{2}{3}, \frac{2}{1}$

Pages 9–10

1 a 6 b $2\frac{3}{8}$ c $3\frac{1}{4}$ d $\frac{7}{8}$ e $4\frac{1}{8}$ f $3\frac{3}{8}$
g $6\frac{3}{8}$ h $5\frac{1}{8}$ i $12\frac{1}{4}$ j 6 k $3\frac{1}{3}$ l $7\frac{5}{6}$
m $6\frac{1}{2}$ n $5\frac{3}{5}$ o $3\frac{3}{10}$ p $3\frac{1}{6}$ q $5\frac{2}{3}$ r $6\frac{2}{9}$

2 a $6\frac{11}{56}, 5\frac{13}{35}, 7\frac{43}{63}, 6\frac{2}{7}, 6\frac{58}{77}$ b $4\frac{5}{24}, 3\frac{23}{60}, 5\frac{25}{36}, 4\frac{25}{84}, 4\frac{101}{132}$
c $5\frac{9}{40}, 4\frac{2}{5}, 6\frac{32}{45}, 5\frac{11}{35}, 5\frac{43}{55}$ d $7, 6\frac{7}{40}, 8\frac{35}{72}, 7\frac{5}{56}, 7\frac{49}{88}$
e $4\frac{11}{24}, 3\frac{19}{30}, 5\frac{17}{18}, 4\frac{23}{42}, 5\frac{1}{66}$ f $5\frac{29}{72}, 4\frac{26}{45}, 6\frac{8}{9}, 5\frac{31}{63}, 5\frac{95}{99}$
g $6\frac{13}{40}, 5\frac{1}{2}, 7\frac{73}{90}, 6\frac{29}{70}, 6\frac{97}{110}$ h $7\frac{7}{24}, 6\frac{7}{15}, 8\frac{7}{9}, 7\frac{8}{21}, 7\frac{28}{33}$

3 a $8\frac{7}{9}$ b $3\frac{23}{60}$

4 a $4\frac{31}{72}$ b $10\frac{7}{12}$ c $5\frac{1}{90}$ d $3\frac{23}{60}$
e $8\frac{29}{56}$ f $5\frac{35}{36}$ g $5\frac{10}{77}$ h $8\frac{11}{12}$
i $3\frac{13}{120}$ j $2\frac{11}{90}$ k $8\frac{11}{28}$ l $3\frac{17}{84}$
m $6\frac{19}{36}$ n $4\frac{23}{24}$ o $8\frac{8}{21}$ p $6\frac{3}{40}$

Pages 11–12

1 a $2\frac{7}{8}$ b $1\frac{3}{4}$ c $2\frac{5}{8}$ d $\frac{7}{8}$ e $2\frac{7}{8}$ f $1\frac{1}{8}$
g $3\frac{5}{8}$ h $4\frac{1}{3}$ i $2\frac{1}{3}$ j $1\frac{5}{6}$ k $2\frac{1}{2}$ l $2\frac{1}{9}$
m $3\frac{1}{9}$ n $2\frac{4}{9}$ o $1\frac{9}{10}$ p $2\frac{3}{10}$ q $\frac{3}{5}$ r $4\frac{4}{5}$
s $1\frac{9}{10}$ t $1\frac{8}{9}$ u $\frac{1}{6}$ v $\frac{11}{12}$ w $2\frac{11}{12}$ x $3\frac{2}{3}$

2 a $4\frac{7}{24}, 2\frac{5}{12}$ b $5\frac{33}{40}, 3\frac{21}{40}$ c $2\frac{27}{28}, 1\frac{3}{14}$
d $4\frac{20}{21}, 2\frac{13}{21}$ e $4\frac{8}{15}, 3\frac{1}{30}$ f $3\frac{19}{20}, 2\frac{47}{60}$
g $1\frac{11}{15}, \frac{59}{60}$ h $5\frac{13}{18}, 2\frac{17}{36}$ i $\frac{61}{24}, \frac{7}{8}$

3 a $2\frac{8}{15}$ b $\frac{55}{56}$ c $2\frac{13}{44}$ d $3\frac{23}{24}$ e $1\frac{11}{14}$ f $2\frac{22}{35}$
g $2\frac{17}{18}$ h $2\frac{59}{60}$ i $3\frac{29}{30}$ j $3\frac{21}{40}$ k $4\frac{13}{36}$ l $3\frac{7}{20}$

Pages 12–13

1 a 3 b 3 c 3 d 4 e 4 f 10
g 3 h 3 i $1\frac{1}{4}$ j $4\frac{1}{3}$ k $2\frac{1}{5}$ l $1\frac{3}{7}$
m $1\frac{1}{7}$ n $2\frac{1}{4}$ o $2\frac{1}{8}$ p $2\frac{1}{7}$

2 a $4\frac{3}{8}, 6\frac{5}{12}, 13\frac{8}{15}, 8\frac{5}{9}, 3\frac{1}{3}$ b $3\frac{9}{32}, 4\frac{13}{16}, 10\frac{3}{20}, 6\frac{5}{12}, 2\frac{1}{2}$
c $15\frac{5}{16}, 22\frac{11}{24}, 47\frac{11}{30}, 29\frac{17}{18}, 11\frac{2}{3}$ d $8\frac{13}{16}, 12\frac{37}{40}, 27\frac{13}{50}, 17\frac{7}{30}, 6\frac{5}{7}$
e $4\frac{29}{64}, 6\frac{17}{32}, 13\frac{31}{40}, 8\frac{17}{24}, 3\frac{11}{28}$ f $2\frac{5}{8}, 3\frac{17}{20}, 8\frac{3}{25}, 5\frac{2}{15}, 2$
g $10\frac{5}{16}, 15\frac{1}{8}, 31\frac{9}{10}, 20\frac{1}{6}, 7\frac{6}{7}$ h $6\frac{27}{28}, 10\frac{3}{14}, 21\frac{19}{35}, 13\frac{13}{21}, 5\frac{15}{49}$

3 a $8\frac{32}{77}, 50\frac{3}{4}, 21\frac{1}{12}, \frac{39}{44}, 28\frac{1}{20}, \frac{13}{16}, 5\frac{34}{35}, 6\frac{1}{4}, 1\frac{1}{10}$
b $2\frac{17}{24}, 11\frac{2}{5}, 4\frac{13}{30}, 4\frac{7}{12}, 4\frac{2}{5}, 58\frac{2}{3}, 2\frac{19}{22}, 4\frac{8}{15}, 1\frac{19}{36}$
c $44\frac{4}{5}, 2\frac{1}{5}, 5\frac{31}{40}, \frac{45}{56}, 23\frac{1}{7}, 18\frac{19}{24}, 4\frac{3}{8}, 4\frac{27}{40}, 4\frac{19}{42}$

Pages 14–15

1 a 7 b 14 c 16 d 33 e 17 f 23
g 16 h 44 i 6 j 18 k 26 l 30
m 8 n 22 o 22 p 46 q 34 r 28
s 33 t 24 u 39 v 15 w 22 x 38

2 a $\frac{13}{40}$ b $6\frac{22}{27}$ c $5\frac{5}{14}$ d $2\frac{2}{5}$ e $1\frac{1}{2}$ f $\frac{3}{8}$
g $7\frac{2}{9}$ h $2\frac{17}{20}$ i $\frac{16}{27}$ j $1\frac{44}{75}$ k $1\frac{83}{117}$ l $4\frac{1}{2}$
m $2\frac{1}{55}$ n $21\frac{5}{16}$ o $1\frac{29}{45}$ p $7\frac{11}{32}$ q $3\frac{6}{13}$ r $1\frac{17}{55}$
s $\frac{52}{63}$ t $2\frac{7}{20}$ u $2\frac{31}{70}$ v $\frac{127}{135}$ w $6\frac{29}{36}$ x $\frac{23}{45}$

3 a b

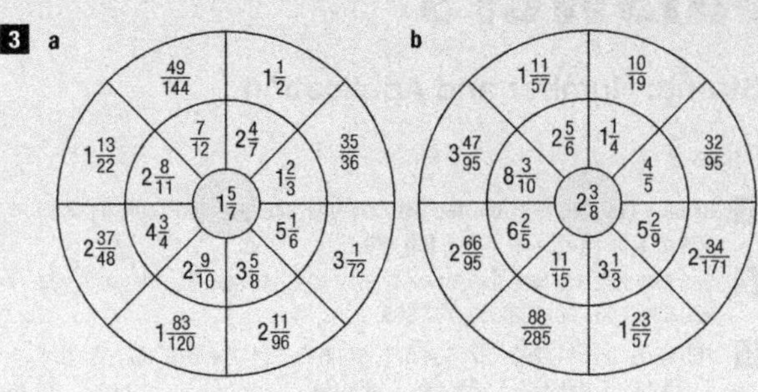

4 a $2\frac{18}{35}$ b $\frac{47}{63}$ c $3\frac{3}{70}$ d $11\frac{11}{15}$ e $2\frac{23}{38}$ f $\frac{65}{108}$
g $\frac{8}{11}$ h $\frac{2}{3}$ i $1\frac{53}{195}$ j $\frac{3}{7}$ k $6\frac{27}{32}$ l $4\frac{4}{39}$
m $\frac{156}{209}$ n $10\frac{17}{20}$ o $\frac{301}{390}$ p $\frac{14}{57}$ q $\frac{19}{32}$ r $\frac{82}{119}$
s $\frac{3}{5}$ t $\frac{27}{55}$

Page 16

1 a $10\frac{1}{2}$ b $3\frac{1}{8}$ c $11\frac{1}{3}$ d $3\frac{3}{8}$ e $18\frac{1}{3}$ f $12\frac{31}{35}$
g $5\frac{25}{36}$ h $4\frac{3}{5}$ i $9\frac{93}{110}$ j $7\frac{13}{40}$ k $2\frac{23}{30}$ l $2\frac{9}{104}$
m $2\frac{17}{40}$ n $3\frac{6}{35}$ o $8\frac{37}{40}$ p $10\frac{13}{30}$ q $2\frac{18}{301}$ r $3\frac{73}{125}$
s $10\frac{83}{100}$ t $3\frac{32}{35}$

2 a 6.4 b 2.45 c 16.5 d 4.32 e 0.125 f 11.424
g 13.81 h 6.26 i 6.818 j 8.905 k 4.092 l 8.5
m 4.43 n 12 o 8.344 p 6.411 q 6.5 r 17.051
s 10.347 t 6.225

3 a 29.59 b 30.32 c $1\frac{53}{84}$ d 8.736 e 27.206 f $3\frac{1}{30}$
g $18\frac{13}{72}$ h 9.235 i 34.915 j 6.514 k $18\frac{8}{15}$ l $8\frac{17}{18}$
m $58\frac{53}{120}$ n 7.443 o $36\frac{2}{3}$ p $9\frac{16}{45}$ q 53.22 r $6\frac{179}{300}$

4 a–b Teacher to check

Pages 17–18

1 $\frac{23}{24}$ kg

2 a 4 tops b $\frac{45}{64}$ m

3 a $\frac{3}{7}$ b 18 c 24 d 27

4 $22\frac{13}{15}$ L **5** $16\frac{13}{20}$ L

6 a 450 b 180 c 600 d 560

7 a $12\frac{1}{12}$ hours b $26\frac{7}{12}$ hours c $67\frac{2}{3}$ hours d $38\frac{2}{3}$ hours

8 $14\frac{7}{10}$ bottles

9 a $23\frac{23}{40}$ kg b $20\frac{33}{40}$ kg

10 $22\frac{14}{15}$ kg **11** 19 scarves and $\frac{16}{21}$ metres left over

12 a 22 b 5 c 7

13 a $14\frac{1}{4}$ km b $8\frac{13}{15}$ km c $4\frac{13}{30}$ km d $24\frac{5}{18}$ km

14 $28\frac{17}{30}$ kg

15 a $3\frac{3}{5}$ m^2 b 6 m^2 c $4\frac{4}{5}$ m^2

16 $17\frac{1}{4}$ metres

17 a $50\frac{39}{40}$ kg b $5\frac{1}{40}$ kg c $2\frac{7}{8}$ kg d $16\frac{119}{120}$ kg

18 a $46\frac{3}{20}$ km b $92\frac{3}{10}$ km

Pages 19–21

1 a Sam's Store = K37.85; Super Mart = K97.60; Save More = K80.35
b K42.50 c K17.25 d K215.80 e K16.65

2 a K3741.90 b More; K220.25 c K3500.45 d K3458.50
e K3805.05 f K151.10

3 a

160.78	188.34	170.4	435.2
36.03	63.59	45.65	310.45
16.148	43.708	25.768	290.568
73.41	100.97	83.03	347.83

b

32.43	20.928	107.1	2.46
314.11	367.46	239.44	349
27.66	81.018	47.01	62.55
50.947	2.411	125.617	16.057

c

117.86	165.316	241.716	140.116
36.131	83.587	159.987	58.387
87.464	134.92	211.32	109.72
375.144	422.6	499	397.4

d

65.21	39.08	11.3	147.31
162.3	188.43	238.81	80.2
6.47	19.66	70.04	88.57
73.526	47.396	2.984	155.626

e

91.107	13.952	73.767	193.967
499.84	422.685	482.5	602.7
182.58	105.425	165.24	285.44
209.64	132.485	192.3	312.5

f

181.71	15.39	12.09	93.63
130.9	66.2	38.72	42.82
207.4	10.3	37.78	119.32
25.95	171.15	143.67	62.13

4 a

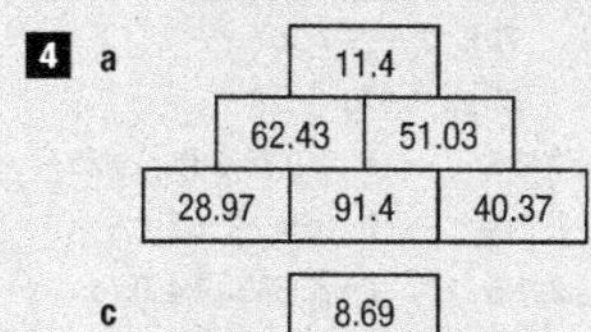

b

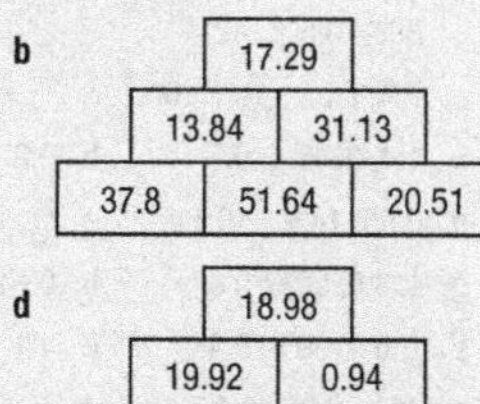

c

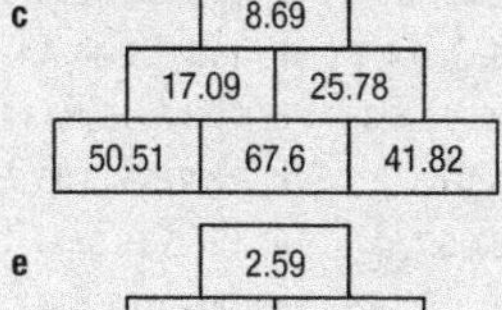

d

18.98
19.92 0.94
52.75 32.83 33.77

e

2.59
21.68 19.09
32.01 53.69 34.6

f

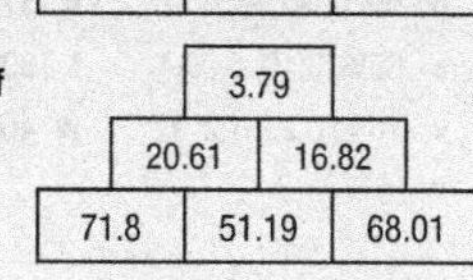

5 a b c d

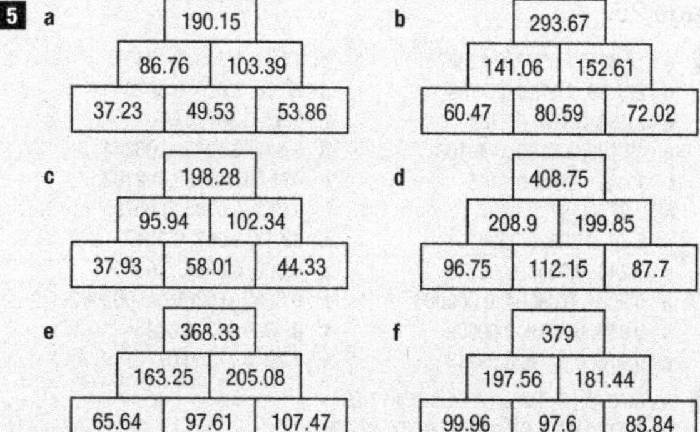

e

368.33
163.25 205.08
65.64 97.61 107.47

f

379
197.56 181.44
99.96 97.6 83.84

Page 22

1 a coffee = K41.75; noodles = K30; rice = K99.60; tea = K27.20; sugar = K27.75; salt = K5.50; powdered milk = K33; sweet biscuits = K47.20; dry biscuits = K42.55; tomato sauce = K25.20
b K120.25

2 a 167.44, 1674.4, 16 744 b 58.17, 581.7, 5817
c 642.8, 6428, 64 280 d 4317.7, 43 177, 431 770
e 68.92, 689.2, 6892 f 31.5, 315, 3150
g 2664, 26 640, 266 400 h 5.73, 57.3, 573
i 190.8, 1908, 19 080 j 310.97, 3109.7, 31 097
k 116, 1160, 11 600 l 58.65, 586.5, 5865
m 242.82, 2428.2, 24 282 n 739, 7390, 73 900
o 447.6, 4476, 44 760 p 1563, 15 630, 156 300
q 90.66, 906.6, 9066 r 751.8, 7518, 75 180
s 37, 370, 3700 t 0.8, 8, 80
u 3439, 34 390, 343 900 v 81.3, 813, 8130

Page 23

1 740.1 **2** 9784.8 **3** 191.06
4 3808 **5** 11 795 **6** 4811.8
7 339.76 **8** 9787.5 **9** 1493.2
10 718.08 **11** 16 435 **12** 2802
13 22 032 **14** 3283.08 **15** 3738
16 11 809 **17** 289.12 **18** 8990
19 4134.27 **20** 6171 **21** 7976.7
22 2919.6 **23** 7092 **24** 16 461

Page 24

1 a 49.706 b 280.747 c 210.04 d 1114.88
e 52.1262 f 449.42

2 a 788.018 b 35.6454 c 85.2992 d 844.936
e 24.6261 f 188.8502

3 a 1361.865 b 1980.72 c 23.0676 d 107.69184
e 1047.826 f 329.4197

4 a 177.5358 b 3845.92 c 530.785 d 229.43635
e 30.17456 f 515.3998

Page 25

1 a 12.367, 1.2367, 0.12367 b 17.43, 1.743, 0.1743
c 277.59, 27.759, 2.7759 d 36.28, 3.628, 0.3628
e 21.54, 2.154, 0.2154 f 8.67, 0.867, 0.0867
g 49.108, 4.9108, 0.49108 h 6.344, 0.6344, 0.06344
i 11.85, 1.185, 0.1185 j 421.63, 42.163, 4.2163
k 2.97, 0.297, 0.0297 l 0.785, 0.0785, 0.00785
m 5.36, 0.536, 0.0536 n 32.17, 3.217, 0.3217
o 0.248, 0.0248, 0.00248 p 8.006, 0.8006, 0.08006
q 0.9004, 0.09004, 0.009004 r 0.3662, 0.03662, 0.003662
s 0.059, 0.0059, 0.00059 t 0.43, 0.043, 0.0043
u 9.218, 0.9218, 0.09218 v 7.72, 0.772, 0.0772

2 a 6.73, 8.16, 5.25, 12.47, 9.88, 11.29
b 4.87, 2.06, 8.54, 15.33, 12.07, 17.23
c 16.46, 9.54, 21.36, 18.97, 11.56, 27.89
d 7.84, 9.38, 12.47, 8.18, 19.58, 16.43

Page 26

1 a 9.35 b 12.6 c 5.68 d 7.27 e 14.7 f 6.58
2 a 15.4 b 11.57 c 9.44 d 15.25 e 18.47 f 23.36
3 a 6.51 b 8.78 c 5.64 d 12.73 e 9.87 f 11.54
4 a 4.36 b 7.08 c 16.4 d 13.18 e 21.56 f 18.37
5 a 9.52 b 9.08 c 14.27 d 23.4 e 17.43 f 19.58
6 a 12.18 b 16.38 c 9.64 d 15.07 e 23.8 f 20.66

Page 27

1 6.7 **2** 9.4 **3** 11.6 **4** 13.4 **5** 17.7
6 18.55 **7** 8.92 **8** 72.86 **9** 26.32 **10** 39.28
11 25.99 **12** 35.5 **13** 35.42 **14** 35.41 **15** 22.91
16 13.65 **17** 23.22 **18** 26.19 **19** 40.21 **20** 20.72

Page 28

All prices have been rounded to the nearest 5 toea.

1 a K33.40 b K16.10 c K21.40
2 a K273.90 b K277.70 c K284 d K283.10
3 9.88 metres
4 a 47 b 33 c 38 d 19
5 a 320.1 km b 38.4 km c 317.8 km
6 253.14 litres
7 a 11.7 kg b 150.33 kg
8 a 36.044 b 0.466 c 0.168

Page 29

1 a $\frac{1}{4}$ b $\frac{4}{5}$ c $\frac{12}{25}$ d $\frac{7}{20}$ e $\frac{4}{25}$ f $\frac{31}{50}$
g $\frac{27}{50}$ h $\frac{3}{25}$ i $\frac{1}{25}$ j $\frac{77}{100}$ k $\frac{17}{20}$ l $\frac{7}{25}$
m $\frac{47}{250}$ n $\frac{51}{200}$ o $\frac{9}{125}$ p $\frac{3}{8}$ q $\frac{5}{8}$ r $\frac{3}{50}$
s $\frac{8}{25}$ t $\frac{2}{5}$ u $\frac{24}{25}$ v $\frac{11}{25}$ w $\frac{33}{50}$ x $\frac{31}{125}$

2 a $2\frac{3}{5}$ b $7\frac{1}{20}$ c $12\frac{19}{50}$ d $5\frac{1}{8}$ e $9\frac{31}{500}$ f $15\frac{3}{4}$
g $8\frac{37}{250}$ h $4\frac{14}{25}$ i $11\frac{3}{500}$ j $25\frac{9}{50}$ k $14\frac{5}{8}$ l $6\frac{58}{125}$
m $10\frac{1}{50}$ n $8\frac{19}{250}$ o $5\frac{7}{8}$ p $3\frac{13}{50}$ q $7\frac{3}{4}$ r $9\frac{433}{500}$
s $4\frac{39}{500}$ t $7\frac{31}{200}$ u $10\frac{21}{50}$ v $8\frac{17}{50}$ w $11\frac{22}{25}$ x $2\frac{64}{125}$

3 a 0.3 b 0.77 c 0.8 d 0.61 e 0.15 f 0.04
g 0.138 h 0.513 i 0.089 j 0.32 k 0.434 l 0.72
m 0.252 n 0.624 o 0.55 p 0.595 q 0.7 r 0.875
s 0.56 t 0.6 u 0.532 v 0.4 w 0.053 x 0.62

4 a 2.6 b 1.9 c 16.19 d 11.3 e 6.75 f 5.125
g 10.58 h 9.52 i 25.13 j 8.4 k 5.63 l 12.371
m 7.3 n 11.145 o 4.224 p 9.22 q 3.76 r 1.009
s 12.7 t 8.298

5 a 0.375 b 0.417 c 0.286 d $0.\dot{7}$ e $0.1\dot{6}$ f $0.\dot{6}$
g $0.\dot{2}\dot{7}$ h 0.857 i 0.083 j $0.8\dot{3}$ k $0.\dot{1}$ l $0.\dot{6}\dot{3}$
m $0.\dot{4}$ n $0.\dot{3}$ o $0.\dot{7}\dot{2}$ p 0.583 q 0.571 r $0.7\dot{3}$

Page 30

1 a 0.6, 60% b $\frac{71}{100}$, 71% c $\frac{49}{100}$, 0.49 d $\frac{21}{50}$, 42%
e 0.125, $12\frac{1}{2}$% f $\frac{7}{25}$, 28% g 0.55, 55% h $\frac{3}{100}$, 3%
i $\frac{54}{125}$, 43.2% j $\frac{53}{100}$, 53% k 0.17, 17% l $\frac{11}{20}$, 55%
m $\frac{2}{25}$, 8% n 0.375, $37\frac{1}{2}$% o $\frac{3}{20}$, 0.15 p $\frac{27}{40}$, $67\frac{1}{2}$%
q 0.7, 70% r $\frac{103}{125}$, 82.4% s $\frac{7}{40}$, 0.175 t 0.42, 42%
u $\frac{4}{25}$, 16% v $\frac{24}{25}$, 0.96 w $\frac{1}{200}$, 0.5% x 0.36, 36%

2 a $1\frac{3}{5}$, 160% b $2\frac{27}{100}$, 2.27 c 1.9, 190% d 3.4, 340%
e $2\frac{3}{4}$, 275% f $1\frac{7}{50}$, 1.14 g $1\frac{3}{1000}$, 100.3% h $5\frac{2}{5}$, 540%
i 2.88, 288% j $\frac{51}{400}$, 0.1275 k $\frac{77}{200}$, 0.385 l $4\frac{7}{100}$, 407%
m 1.85, 185% n $1\frac{14}{25}$, 1.56 o 2.34, 234% p $3\frac{257}{500}$, 351.4%
q $\frac{29}{400}$, 0.0725 r $1\frac{12}{25}$, 148% s 3.8, 380% t $2\frac{183}{1000}$, 218.3%
u 5.5, 550% v $\frac{47}{200}$, 0.235 w $5\frac{11}{50}$, 522% x 7.2, 720%

Page 31

1 a $\frac{3}{5}$ b $\frac{3}{4}$ c $\frac{2}{3}$ d $1\frac{1}{2}$ e $1\frac{2}{3}$ f $\frac{1}{3}$
g 3 h $\frac{5}{6}$ i $\frac{1}{2}$ j $2\frac{2}{3}$ k $2\frac{1}{5}$ l $\frac{2}{7}$
m $\frac{2}{5}$ n $1\frac{1}{3}$ o $1\frac{1}{8}$ p $\frac{3}{5}$ q $\frac{4}{5}$ r $\frac{5}{8}$
s $2\frac{1}{3}$ t $1\frac{4}{9}$ u $\frac{2}{15}$ v $1\frac{3}{7}$ w $\frac{2}{3}$ x $\frac{3}{5}$

2 a 2:5 b 7:4 c 2:9 d 2:3 e 12:5 f 2:3
g 11:8 h 18:7 i 3:7 j 15:8 k 3:7 l 3:11
m 13:4 n 17:2 o 1:5 p 19:7 q 3:10 r 10:7
s 1:8 t 16:3 u 5:3 v 4:9 w 3:10 x 8:17

3 a
35% 0.35 $\frac{7}{20}$ 7:20
90% 0.9 $\frac{9}{10}$ 9:10
16% 0.16 $\frac{4}{25}$ 4:25
80% 0.8 $\frac{4}{5}$ 4:5
7% 0.07 $\frac{7}{100}$ 7:100

b
12% 0.12 $\frac{3}{25}$ 3:25
75% 0.75 $\frac{3}{4}$ 3:4
48% 0.48 $\frac{12}{25}$ 12:25
70% 0.7 $\frac{7}{10}$ 7:10
$37\frac{1}{2}$% 0.375 $\frac{3}{8}$ 3:8

4 a 1:4, 33%, $\frac{1}{2}$, 0.7 b 0.3, 55%, $\frac{4}{5}$, 7:5 c $\frac{3}{4}$, 0.76, 83%, 24:25
d 0.05, 14%, $\frac{1}{3}$, 6:11 e $\frac{21}{50}$, 1:2, 0.7, 71% f 85%, $\frac{9}{10}$, 1.2, 8:5
g $\frac{1}{8}$, 3:8, 50%, 0.8 h 23%, $\frac{2}{3}$, 3:2, 2.5 i $\frac{3}{10}$, 36%, 2:4, 0.75
j $\frac{1}{9}$, 0.9, 99%, 9:4 k 4%, 1:20, $\frac{1}{10}$, 0.4 l 14%, $\frac{1}{4}$, 0.44, 4:1
m 18%, 0.35, $\frac{3}{5}$, 4:5 n 57%, 1.25, $1\frac{1}{2}$, 5:3 o $1\frac{1}{4}$, 1.3, 150%, 3:1
p 5%, 2:25, $\frac{4}{25}$, 0.17 q 3:4, 81%, 0.83, $\frac{9}{10}$ r $2\frac{1}{2}$, 275%, 3.05, 7:2
s 29%, 3:10, $\frac{7}{20}$, 0.4 t 60%, 0.63, 5:7, $\frac{9}{10}$ u 1:5, 0.34, 37%, $\frac{21}{50}$
v 18%, 0.25, 1:3, $\frac{4}{9}$ w 4:7, 100%, $1\frac{3}{5}$, 1.75 x 78%, 0.87, $\frac{7}{8}$, 8:7

Page 32

1 a 21.6% b 18.6% c 20.5% d 22.7% e 10.4% f 16.2% g 37.9% h 12.2% i 14.9% j 15.8% k 14.4% l 14.1% m 19.6% n 14.8% o 4.8% p 5.3%

2 a 60% b 8% c 76.8% d 24.3% e 30.4% f 58.3% g 5.4% h 23% i 22.5% j 26.9% k 32.1% l 22.3% m 18.6% n 36.5% o 10.2%

3 a 32.9% b 45.3% c 30.5% d 18.7% e 59.1% f 19.7% g 47.6% h 65.4% i 60.3% j 63.8% k 42.5% l 40.5% m 60.1% n 48.9% o 30.1%

4 Wife: 61.5%; Daughter: 16.4%; Grandsons: 3.6% each; Brother: 11.3%

Page 33

1 a 1.15 b 1.2 c 1.1 d 1.3 e 1.12 f 1.25 g 1.5 h 1.05 i 1.08 j 1.095 k 1.225 l 1.065 m 1.112 n 1.155 o 1.027 p 1.268 q 1.124 r 1.205

2 a 0.9 b 0.5 c 0.95 d 0.65 e 0.82 f 0.85 g 0.8 h 0.7 i 0.6 j 0.25 k 0.93 l 0.75 m 0.875 n 0.945 o 0.965 p 0.825 q 0.918 r 0.833

3 a 62.5 b 26.4 c 15.6 d 37.8 e 32.4 f 72.45 g 20.43 h 50.63 i 9.9 j 59.79 k 22.05 l 83.3

4 a 34 b 34.2 c 44.8 d 18.2 e 17.1 f 19.8 g 29.05 h 48 i 28.16 j 47 k 36.19 l 63.75

5 a 120.75 b 287.5 c 147.2 d 358.8 e 271.4 f 529 g 396.75 h 211.6 i 241.5 j 546.25 k 583.05 l 443.9 m 318.55 n 469.2 o 673.9 p 186.3 q 378.35 r 342.7

6 a 162.75 b 297.6 c 228.78 d 351.54 e 169.26 f 479.88 g 401.76 h 101.37 i 330.15 j 269.7 k 529.17 l 131.13 m 236.22 n 481.74 o 151.59 p 453.84 q 371.07 r 219.48

Page 34

1 a 12.5% b 21.9% c 23.1% d 42.9% e 25% f 52.9% g 65.2% h 42.9% i 21.4% j 37.5% k 41.9% l 14.9% m 29.3% n 28.9% o 24.7%

2 a 25.9% b 37.5% c 20% d 26.8% e 27.8% f 30% g 12.5% h 37.5% i 8.3% j 10.5% k 23.6% l 9.6% m 24.5% n 12.5% o 20%

3 a 42.9% b 6.25% c 20.3% d 12.5% e 29.1% f 3.2% g 11.2% h 10.5% i 15.2% j 24% k 6.3% l 13.3%

4 a 32.1% b 11.8% c 5.6% d 4.6% e 6.6% f 8% g 6.7% h 9.7% i 14.2% j 12.3% k 8.6% l 10%

Page 35

1 a K258.50 b K139.70 c K393.80 d K160.60 e K301.40 f K498.30 g K136.95 h K102.85 i K567.60 j K309.10 k K83.05 l K339.90 m K470.80 n K402.60 o K200.75 p K242.55 q K190.85 r K269.50

2 a K41.60 b K21.10 c K74.90 d K201.80 e K138.20 f K288.70 g K570.60 h K364.90 i K281.70 j K729.40 k K655.20 l K391.80 m K50.65 n K37.35 o K98.85 p K23.95 q K11.75 r K46.55

3 a K5.80 b K7.60 c K4.10 d K3.80 e K3.20 f K5.20 g K9.60 h K8.50 i K16.40 j K15.40 k K13.60 l K12.10 m K19.80 n K16.30 o K20.40 p K31.20 q K19.70 r K23.50 s K36.40 t K32.10 u K30.90 v K41.50 w K17.30 x K25.90

4 a K43 b K28 c K66 d K57 e K39 f K82 g K97 h K153 i K133 j K171 k K125 l K158 m K205 n K189 o K216 p K254 q K286 r K248 s K162 t K311 u K418 v K472 w K443 x K307

Page 36

1 a K38.25 b K99 c K71.40 d K16.15 e K116 f K390.60 g K208.25 h K70 i K510.75 j K141 k K54.40 l K146.30 m K250 n K122.10 o K287.70 p K44.55 q K225.75 r K658.75 s K365.20 t K67.55 u K255.30 v K321.30

2 a K2415 b K1551.50 c K2062.50 d K5670 e K2903.25 f K3527.75 g K2139.20 h K2722.50 i K5160 j K3692 k K3210.05 l K1160.85 m K7284.60 n K4620 o K3381 p K8062.50 q K2819.60 r K4284.50 s K8967.50 t K6515.70 u K6880.10 v K5164.65

Page 37

1 a K75 b K162.50 c K56.25 d K187.50 e K112.50 f K31.25 g K218.75 h K325 i K137.50 j K281.25 k K400 l K287.50 m K155 n K385 o K320 p K222.50 q K370 r K180

2 a K1200 b K900 c K2100 d K1800 e K2600 f K4200 g K3700 h K3500 i K5300 j K4700 k K3900 l K5600 m K7300 n K9100 o K8000 p K6800 q K8900 r K9800

3 a Done
b K1200, K420, K180, K720, K900, K540
c K1700, K595, K255, K1020, K1275, K765
d K3400, K1190, K510, K2040, K2550, K1530
e K2500, K875, K375, K1500, K1875, K1125
f K4100, K1435, K615, K2460, K3075, K1845
g K3700, K1295, K555, K2220, K2775, K1665
h K3550, K1242.50, K532.50, K2130, K2662.50, K1597.50
i K4250, K1487.50, K637.50, K2550, K3187.50, K1912.50
j K6120, K2142, K918, K3672, K4590, K2754
k K5640, K1974, K846, K3384, K4230, K2538
l K2830, K990.50, K424.50, K1698, K2122.50, K1273.50
m K1470, K514.50, K220.50, K882, K1102.50, K661.50
n K3280, K1148, K492, K1968, K2460, K1476

Pages 38–39

1 K1840

2 a 910 litres b 490 litres

3 a K125.90 b K1384.90

4 K2371.60

5 a 722 b 114 c 418 d 836 e 684

6 40.9% **7** K236.50 **8** K166.75

9 K17 300 **10** 3.75% **11** 6.47%

12 a K5380 b K269 c K7649

13 a K33 300 b K34 225 c K35 150 d K31 450 e K32 560 f K34 040

14 72 males

15 a K329 b K3290

16 a 58.2% b 41.8%

17 a K43.75 b K105 c K70 d K175
e K131.25 f K157.50

18 a $12\frac{1}{2}$% off K80 is cheaper b K2.25

19 K625 **20** 41.8% **21** 216 members

Pages 40–41

1 a 1:4 b 1:9 c 1:8 d 11:1 e 1:14 f 5:4
g 2:3 h 3:2 i 5:12 j 4:3 k 5:9 l 3:4
m 4:9 n 14:9 o 2:1 p 9:2 q 5:19 r 4:21

2 a 1:300 b 20:3 c 1:20 d 75:4 e 40:3 f 1:9
g 3:20 h 100:1 i 1:8 j 100:9 k 1:10 l 1:125
m 1:25 n 45:4 o 100:1 p 5:2 q 5:12 r 3:16

3 a 3:5 b 1:2 c 6:1 d 6:5 e 7:9 f 10:3
g 23:15 h 16:29 i 6:7 j 16:15 k 5:6 l 7:4
m 13:7 n 1:4 o 4:1 p 14:1 q 1:10 r 20:1
s 1:4 t 25:21 u 1:10 v 15:1 w 1:3 x 15:1

Page 41

1 a 4:12 b 21:27 c 16:12 d 10:15 e 30:25 f 9:15
g 12:42 h 45:10 i 7:56 j 36:24 k 21:12 l 16:40
m 18:8 n 20:32 o 36:42 p 44:28

2 a 3:4 b 7:3 c 8:3 d 5:8 e 2:5 f 2:5
g 3:2 h 1:3 i 5:8 j 4:7 k 6:7 l 7:10
m 3:1 n 2:3 o 16:9 p 19:12

3 a 9:2 b 33:44 c 54:48 d 4:9 e 21:35 f 3:4
g 2:5 h 42:72 i 32:12 j 13:3 k 7:4 l 6:11
m 2:5 n 20:52 o 45:30 p 10:9

4 a 18:30:30 b 12:30:42 c 2:4:3 d 25:5:15
e 2:1:3 f 24:15:9 g 3:2:3 h 3:5:1
i 15:35:20 j 12:20:28 k 3:5:4 l 21:6:30
m 8:5:7 n 35:14:21 o 36:63:36

Page 42

1 a $\frac{2}{5}, \frac{3}{5}$ b $\frac{4}{9}, \frac{5}{9}$ c $\frac{1}{4}, \frac{3}{4}$ d $\frac{5}{7}, \frac{2}{7}$
e $\frac{7}{9}, \frac{2}{9}$ f $\frac{1}{7}, \frac{6}{7}$ g $\frac{3}{11}, \frac{8}{11}$ h $\frac{4}{5}, \frac{1}{5}$
i $\frac{4}{6}, \frac{2}{6}$ j $\frac{5}{12}, \frac{7}{12}$ k $\frac{3}{8}, \frac{5}{8}$ l $\frac{8}{10}, \frac{2}{10}$
m $\frac{2}{10}, \frac{7}{10}, \frac{1}{10}$ n $\frac{6}{14}, \frac{5}{14}, \frac{3}{14}$ o $\frac{4}{14}, \frac{3}{14}, \frac{7}{14}$ p $\frac{5}{14}, \frac{1}{14}, \frac{8}{14}$
q $\frac{1}{7}, \frac{4}{7}, \frac{2}{7}$ r $\frac{3}{11}, \frac{4}{11}, \frac{4}{11}$

2 a K20, K60 b K50, K30 c K48, K32 d K24, K56
e K72, K8 f K16, K64 g K10, K70 h K35, K45
i K68, K12 j K25, K55 k K44, K36 l K26, K54
m K25, K40, K15 n K36, K4, K40 o K16, K24, K40 p K28, K20, K32
q K40, K10, K30 r K38, K16, K26

3 a 100 mL, 500 mL b 250 mL, 350 mL c 180 mL, 420 mL
d 200 mL, 400 mL e 225 mL, 375 mL f 320 mL, 280 mL
g 440 mL, 160 mL h 384 mL, 216 mL i 570 mL, 30 mL
j 340 mL, 260 mL k 125 mL, 475 mL l 345 mL, 255 mL
m 160 mL, 280 mL, 160 mL n 360 mL, 60 mL, 180 mL
o 75 mL, 300 mL, 225 mL p 210 mL, 150 mL, 240 mL
q 200 mL, 150 mL, 250 mL r 210 mL, 120 mL, 270 mL

4 a 225 g, 135 g b 288 g, 72 g c 120 g, 240 g d 90 g, 270 g
e 210 g, 150 g f 240 g, 120 g g 120 g, 240 g h 144 g, 216 g
i 280 g, 80 g j 108 g, 252 g k 324 g, 36 g l 45 g, 315 g
m 192 g, 96 g, 72 g n 150 g, 60 g, 150 g o 120 g, 80 g, 160 g
p 36 g, 198 g, 126 g q 150 g, 90 g, 120 g r 200 g, 40 g, 120 g

5 a 144 cm, 96 cm b 24 cm, 216 cm c 80 cm, 160 cm
d 200 cm, 40 cm e 80 cm, 160 cm f 135 cm, 105 cm
g 64 cm, 176 cm h 60 cm, 180 cm i 140 cm, 100 cm
j 84 cm, 156 cm k 165 cm, 75 cm l 190 cm, 50 cm
m 20 cm, 140 cm, 80 cm n 120 cm, 45 cm, 75 cm
o 50 cm, 110 cm, 80 cm p 144 cm, 16 cm, 80 cm
q 45 cm, 30 cm, 165 cm r 60 cm, 12 cm, 168 cm

6 a 160 L, 1040 L b 700 L, 500 L c 1080 L, 120 L d 384 L, 816 L
e 780 L, 420 L f 550 L, 650 L g 1125 L, 75 L h 450 L, 750 L
i 660 L, 540 L j 672 L, 528 L k 525 L, 675 L l 60 L, 1140 L
m 144 L, 816 L, 240 L n 240 L, 760 L, 200 L o 100 L, 50 L, 1050 L
p 675 L, 225 L, 300 L q 480 L, 300 L, 420 L r 500 L, 350 L, 350 L

Pages 43–44

1 a 12 buckets b 20 buckets

2 15 tins

3 a $1\frac{1}{2}$ cups b 6 cups

4 a 25 m and 35 m b 70 m

5 a 252 students b 6 teachers

6 2.4 L of water and 2 L of powder

7 a 400 mL b 1600 mL c 960 mL d 640 mL

8 a 1250 mL b 320 mL
c 6 L of water and 2.4 L of concentrate

9 a 2 cups b 3 cups c $2\frac{1}{2}$ cups

10 a K17 b K12.50 c K23 d K9.50

11 24 L, 60 L and 36 L

12 a 5055 adults b 3370 children

13 a 48 students b 15 adults

14 32 metres

15 a 250 mL b 175 mL c 400 mL

16 a 240 g b 645 g c 1035 g of beef and 345 g of pork

17 a $2\frac{1}{2}$ m b 50 cm c 3.75 m

18 a 48 b 84 coins

19 225 girls

20 a 10 cm b 4 cm c 40 cm^2

Page 45

1 a money and time b money and length c speed and time
d money and weight e capacity and time f pages and time
g bread rolls and time h length and time i money and time
j speed and time k heart beats and time l capacity and area

2 a 840 drinks b K102 c 450 litres d 756 seedlings
e 30 km f 1575 g g 182 words

3 a K9 b K22 c 11.5 km d 15 litres
e 719 litres f 80 kg g $15\frac{5}{8}$ minutes

Page 46

1 8 metres

2 **a** 6 hours **b** 5 hours **c** 4 hours

3 80 hours **4** 253.5 km

5 **a** 1.6 litres ($1\frac{3}{5}$ L) per minute **b** 26.7 mL ($26\frac{2}{3}$ L) per second

6 50 minutes **7** 33 km

8 **a** 3080 km **b** 2420 km **c** 4840 km **d** 660 km

9 525 cattle **10** $3\frac{1}{2}$ hours

11 **a** K962.50 **b** K660 **c** K440 **d** K1292.50
e K2337.50 **f** K1650

12 54 hamburgers

Page 47

1 **a** 750 g at K7.45 **b** 300 g at K15.60 **c** 1 kg at K8.45
d 5 kg at K11.45 **e** 1 L at K8.15 **f** 2.5 kg at K8.65

2 **a** Company A or C **b** Company B **c** Company C
d Company C **e** Company C

3 **a** Company A **b** Company A or B **c** Company C
d Company C **e** Company C

4 **a** Hotel B – K433 **b** Hotel A or B – K325 **c** Hotel C – K370
d Hotel A or C – K305 **e** Hotel C – K715

5 **a** Hotel D – K430.35 **b** Hotel D – K321.10 **c** Hotel C – K370
d Hotel A or C – K305 **e** Hotel C – K715

Page 48

1 **a** $<$ **b** $<$ **c** $>$ **d** $>$ **e** $<$ **f** $<$
g $>$ **h** $<$ **i** $>$ **j** $<$ **k** $>$ **l** $<$
m $<$ **n** $>$ **o** $>$ **p** $<$ **q** $>$ **r** $>$
s $>$ **t** $<$

2 **a** 27 **b** +16 **c** +31 **d** 89 **e** 54 **f** −17
g 49 **h** −52 **i** 19 **j** +7 **k** +45 **l** −19
m 77 **n** 8 **o** 105 **p** −89 **q** +32 **r** 19

3 **a** −36, −27, −8, −4, 11, 16, +18, +25
b −51, −32, −19, −7, 17, +21, 23, +36
c −76, −54, −32, −29, −18, −15, −3, 18
d −33, −23, −11, 35, +40, +61, +70, 89
e −42, −27, −22, +21, 28, 38, +52, +56
f −56, −14, −13, −6, 17, 45, 55, +60
g −57, −50, −44, −39, 0, +9, +44, 77
h −73, −64, −53, −48, 78, +80, 81, +86
i −90, −87, −80, −66, 77, +80, 88, +91
j −60, −54, −47, −21, +12, +47, 54, 60
k −38, −37, −30, +30, +35, 40, +41, 42
l −27, −21, −19, −18, −17, +15, +16, 23
m −15, −14, −10, −7, 8, 9, +11, +16
n −71, −70, −17, 0, +1, 63, +70, +76
o −119, −112, −105, 108, +118, +125
p −143, −140, −134, 130, 137, +139
q −220, −219, −217, 210, +214, 216
r −186, −178, −175, +170, 187, +189
s −161, −160, −158, −155, +154, 164
t −287, −285, −280, 278, +286, 290
u −240, −204, −200, 242, 243, +247
v −321, −315, −312, −310, +312, 315

Page 49

1 **a** = **b** = **c** ≠ **d** = **e** ≠ **f** =
g ≠ **h** = **i** = **j** ≠ **k** = **l** =
m ≠ **n** ≠ **o** = **p** = **q** ≠ **r** =
s ≠ **t** =

2 **a** 1 **b** 4 **c** −1 **d** 17 **e** 9 **f** −15
g 16 **h** −20 **i** 8 **j** 27 **k** −2 **l** −11
m −2 **n** 24 **o** 11 **p** −26 **q** 12 **r** −19
s 30 **t** 9

3 **a** 21 **b** −13 **c** −3 **d** 6 **e** −15 **f** 16
g −5 **h** 13 **i** 13 **j** 0 **k** 11 **l** 3
m 14 **n** 7 **o** 3 **p** 37 **q** 2 **r** 11
s −34 **t** −3

4 **a** −2 **b** −5 **c** −5 **d** 7 **e** −3 **f** −6
g 10 **h** 16 **i** −8 **j** −17 **k** −1 **l** −11
m 12 **n** 24 **o** 30 **p** −3 **q** −8 **r** −10
s −10 **t** −5

Page 50

1 **a**

9	−15	6	21	−12	−33
−18	30	−12	−42	24	66
15	−25	10	35	−20	−55
12	−20	8	28	−16	−44
−24	40	−16	−56	32	88

b

−12	8	−4	16	10	−18
−54	36	−18	72	45	−81
60	−40	20	−80	−50	90
42	−28	14	−56	−35	63
−72	48	−24	96	60	−108

2 **a** −2 **b** −5 **c** 4 **d** 6 **e** −3 **f** 9
g −5 **h** −8 **i** −3 **j** −4 **k** −5 **l** 7
m 7 **n** 8 **o** −8 **p** −4 **q** −5 **r** 5
s −7 **t** 2

3 **a** 36 **b** 96 **c** 120 **d** 70 **e** −120 **f** 108
g −84 **h** −72 **i** −160 **j** −125 **k** −72 **l** 44
m 243 **n** 60 **o** −48 **p** −350 **q** 0 **r** 165
s 192 **t** −112

Page 51

1 **a** −10 **b** 10 **c** −15 **d** −4 **e** −27 **f** 2
g 9 **h** −10 **i** −36 **j** −63 **k** 2 **l** −11
m 2 **n** −56 **o** 2

2 **a**

−19	1	−13	8	−29	−16
12	32	18	39	2	15
−20	0	−14	7	−30	−17
7	27	13	34	−3	10
−42	−22	−36	−15	−52	−39
33	53	39	60	23	36

b

43	18	11	36	5	41
6	−19	−26	1	−32	4
1	−24	−31	−6	−37	−1
68	43	36	61	30	66
−11	−36	−43	−18	−49	−13
3	−22	−29	−4	−35	1

3 **a** **b**

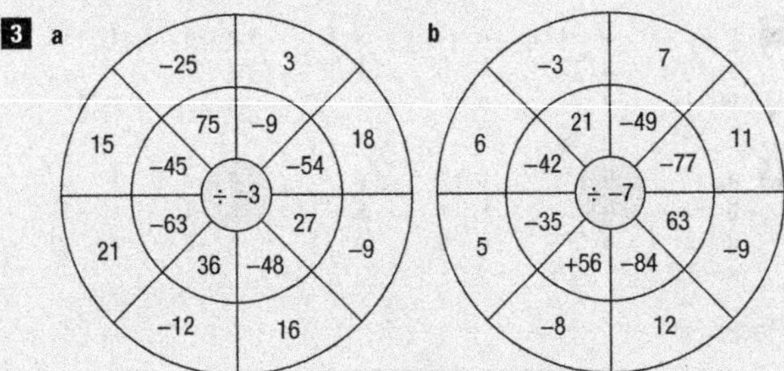

Page 52

1

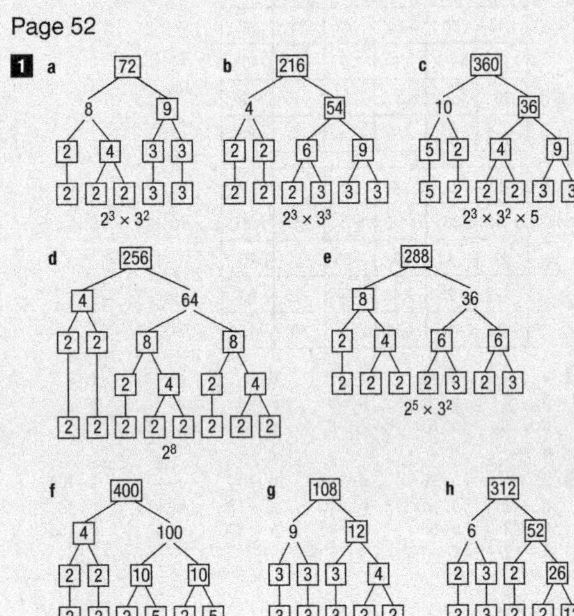

2 **a** 72 **b** 64, $2^2 \times 4^2$
c $2 \times 2 \times 2 \times 2 \times 2 \times 3$, $2^5 \times 3$ **d** 432, $2 \times 2 \times 2 \times 2 \times 3 \times 3 \times 3$
e $3 \times 3 \times 7 \times 7$, $3^2 \times 7^2$ **f** 200, $2 \times 2 \times 2 \times 5 \times 5$
g 784, $2^4 \times 7^2$ **h** $2 \times 2 \times 2 \times 3 \times 5 \times 5$, $2^3 \times 3 \times 5^2$

Pages 53–54

1 **a** 25 **b** 64 **c** 121 **d** 8 **e** 49 **f** 27
g 1000 **h** 225 **i** 125 **j** 400 **k** 729 **l** 144
m 80 **n** 135 **o** 150 **p** 287 **q** 1380 **r** 57
s 72 **t** 64 000 **u** 8 **v** 3267 **w** 256 **x** 5

2 **a** 5 **b** 10 **c** 16 **d** 11 **e** 25 **f** 7
g 18 **h** 30 **i** 14 **j** 9 **k** 12 **l** 32
m 6 **n** 3 **o** 15 **p** 17 **q** 20 **r** 28

3 **a** 3 **b** 10 **c** 5 **d** 11 **e** 20 **f** 4
g 7 **h** 9 **i** 6 **j** 12 **k** 15 **l** 8
m 2 **n** 13 **o** 16 **p** 14 **q** 19 **r** 17

4 **a** 2^3 **b** -4^2 **c** $\sqrt[3]{1331}$ **d** $\sqrt[3]{1000}$ **e** 7^2 **f** 9^2
g -9^2 **h** $\sqrt{64}$ **i** 3^3 **j** $\sqrt{16}$ **k** $\sqrt{144}$ **l** 3^3
m $\sqrt{121}$ **n** 8^3 **o** $\sqrt{49}$ **p** 4^2

5 **a** = **b** > **c** = **d** = **e** < **f** <
g > **h** > **i** > **j** = **k** > **l** <
m > **n** < **o** = **p** <

Pages 54–55

1 **a** 2^7 **b** 8^5 **c** 4^2 **d** 11^3 **e** 6^4 **f** 8^5
g 5^7 **h** 15^2 **i** 3^9

2

$2^0 = 1$	$2^1 = 2$	$2^2 = 4$	$2^3 = 8$	$2^4 = 16$	$2^5 = 32$
$3^0 = 1$	$3^1 = 3$	$3^2 = 9$	$3^3 = 27$	$3^4 = 81$	$3^5 = 243$
$4^0 = 1$	$4^1 = 4$	$4^2 = 16$	$4^3 = 64$	$4^4 = 256$	$4^5 = 1024$
$5^0 = 1$	$5^1 = 5$	$5^2 = 25$	$5^3 = 125$	$5^4 = 625$	$5^5 = 3125$
$6^0 = 1$	$6^1 = 6$	$6^2 = 36$	$6^3 = 216$	$6^4 = 1296$	$6^5 = 7776$
$7^0 = 1$	$7^1 = 7$	$7^2 = 49$	$7^3 = 343$	$7^4 = 2401$	$7^5 = 16\,807$
$8^0 = 1$	$8^1 = 8$	$8^2 = 64$	$8^3 = 512$	$8^4 = 4096$	$8^5 = 32\,768$
$9^0 = 1$	$9^1 = 9$	$9^2 = 81$	$9^3 = 729$	$9^4 = 6561$	$9^5 = 59\,049$
$10^0 = 1$	$10^1 = 10$	$10^2 = 100$	$10^3 = 1000$	$10^4 = 10\,000$	$10^5 = 100\,000$

3 **a** = **b** > **c** > **d** < **e** = **f** >
g = **h** > **i** > **j** > **k** < **l** =
m > **n** = **o** > **p** <

4 **a** 12 **b** 7 **c** 2 **d** 90 **e** 6 **f** 0
g 3 **h** 2 **i** 26 **j** 0 **k** −6 **l** 44
m 500 **n** 4 **o** 2 **p** 0

5 **a** 5 **b** 36 **c** 49 **d** 32 **e** 64 **f** 1
g 5 **h** 27 **i** 2 **j** 100 **k** 5 **l** 8

Pages 55–56

1 **a** 0.2 **b** 0.04
c $\frac{1}{4}$ or 0.25 **d** $1 \div 2 \div 2 \div 2 = \frac{1}{8}$ or 0.125
e $1 \div 4 \div 4 = \frac{1}{16}$ or 0.0625 **f** $1 \div 10 = \frac{1}{10}$ or 0.1
g $1 \div 10 \div 10 = \frac{1}{100}$ or 0.01 **h** $1 \div 10 \div 10 \div 10 = \frac{1}{1000}$ or 0.001
i $1 \div 3 = \frac{1}{3}$ or 0.33 **j** $1 \div 3 \div 3 = \frac{1}{9}$ or 0.11
k $1 \div 8 \div 8 = \frac{1}{64}$ or 0.0156 **l** $1 \div 16 = \frac{1}{16}$ or 0.0625

2 **a** 0.008 **b** 0.111 **c** 0.037 **d** 0.083 **e** 0.004 **f** 0.008
g 0.004 **h** 0.012 **i** 0.05 **j** 0.063 **k** 0.056 **l** 0.016

3 a $\frac{1}{5^2}$ b $\frac{1}{4^3}$ c $\frac{1}{7^2}$ d $\frac{1}{10^4}$ e $\frac{1}{3^5}$ f $\frac{1}{2^8}$
g $\frac{1}{8^1}$ h $\frac{1}{9^4}$ i $\frac{1}{6^2}$ j $\frac{1}{5^5}$ k $\frac{1}{11^2}$ l $\frac{1}{7^3}$
m $\frac{1}{12^4}$ n $\frac{1}{3^{10}}$ o $\frac{1}{8^7}$ p $\frac{1}{4^5}$ q $\frac{1}{10^9}$ r $\frac{1}{15^4}$

4 a $\frac{1}{10^3} = \frac{1}{1000} = 0.001$ b $\frac{1}{4^2} = \frac{1}{16} = 0.063$ c $\frac{1}{6^1} = \frac{1}{6} = 0.167$
d $\frac{1}{2^4} = \frac{1}{16} = 0.063$ e $\frac{1}{3^3} = \frac{1}{27} = 0.037$ f $\frac{1}{8^2} = \frac{1}{64} = 0.016$
g $\frac{1}{12^2} = \frac{1}{144} = 0.007$ h $\frac{1}{5^2} = \frac{1}{25} = 0.04$ i $\frac{1}{2^5} = \frac{1}{32} = 0.031$
j $\frac{1}{4^3} = \frac{1}{64} = 0.016$ k $\frac{1}{7^3} = \frac{1}{343} = 0.003$ l $\frac{1}{11^2} = \frac{1}{121} = 0.008$

5 a 7^{-5} b 12^{-4} c 9^{-5} d 8^{-7} e 11^{-4} f 5^{-6}
g 3^{-8} h 2^{-10} i 12^{-5} j 4^{-4} k 6^{-3} l 10^{-7}
m 5^{-9} n 1^{-5} o 11^{-8} p 9^{-4} q 15^{-6} r 7^{-10}

Page 57 Assessment

1 a < b > c = d > e = f <
g > h <

2 a $\frac{3}{4}, \frac{8}{8}, 1\frac{3}{4}, 1\frac{7}{8}, \frac{17}{8}$ b $\frac{1}{6}, \frac{2}{3}, \frac{5}{6}, 1\frac{1}{3}, \frac{3}{2}$ c $\frac{3}{10}, \frac{1}{2}, \frac{4}{5}, 1\frac{2}{5}, \frac{9}{5}$
d $\frac{1}{4}, \frac{7}{12}, \frac{2}{3}, 1\frac{1}{12}, \frac{12}{7}$ e $\frac{1}{3}, \frac{5}{6}, \frac{6}{5}, \frac{11}{6}, 2\frac{1}{2}$ f $\frac{1}{9}, \frac{2}{3}, \frac{7}{9}, 1\frac{2}{3}, \frac{18}{9}$

3 a $1\frac{7}{8}$ b $1\frac{3}{4}$ c $3\frac{1}{10}$ d $\frac{2}{3}$

4 a $4\frac{2}{21}$ b $4\frac{19}{40}$ c $4\frac{13}{20}$ d $6\frac{11}{24}$ e $4\frac{11}{12}$ f $8\frac{7}{18}$
g $2\frac{1}{12}$ h $1\frac{13}{21}$ i $2\frac{23}{40}$ j $\frac{23}{24}$ k $1\frac{11}{20}$ l $2\frac{29}{36}$

5 a $2\frac{2}{5}$ b $1\frac{27}{28}$ c $7\frac{1}{14}$ d $1\frac{37}{96}$ e 34 f $3\frac{11}{36}$
g $3\frac{7}{27}$ h $3\frac{7}{36}$ i $\frac{117}{140}$ j $\frac{19}{28}$ k $4\frac{14}{19}$ l $3\frac{3}{20}$

6 a $7\frac{7}{10}$ b $2\frac{9}{10}$ c $3\frac{43}{60}$ d $4\frac{23}{28}$ e $3\frac{37}{200}$ f $6\frac{47}{600}$
g $4\frac{463}{900}$ h $3\frac{53}{75}$

7 a $10\frac{26}{45}$ containers b 115 kg c $1\frac{39}{56}$ metres

Page 58 Assessment

1 a < b > c < d < e > f < g < h <

2 a

83.13	102.02	107.47	238.27
15.246	34.136	39.586	170.386
44.64	63.53	68.98	199.78
290.96	309.85	315.3	446.1

b

3.18	59.845	148.7	19.62
199.22	255.885	47.34	215.66
12.86	69.525	139.02	29.3
60.519	3.854	212.399	44.079

3 a 6380 b 90.44 c 2539.6 d 46 134
e 2953.8 f 162.081 g 441.41 h 665.42
i 537.806 j 2399.544 k 1269.315 l 147.645

4 a 8.327 b 4.8335 c 16.56 d 2.141 e 103.3 f 33.54
g 25.8 h 37.22 i 42.7 j 66.8 k 47.3 l 56.8

5 a 463.75 litres b K45.15 c K6.65 d 7.66 metres
e 198.55 km

Page 59 Assessment

1 a $3\frac{9}{10}$ b $\frac{67}{100}$ c $2\frac{4}{5}$ d $5\frac{7}{20}$ e $\frac{3}{50}$ f $1\frac{9}{125}$
g $\frac{133}{500}$ h $8\frac{19}{50}$ i $11\frac{31}{50}$ j $\frac{91}{200}$ k $6\frac{27}{50}$ l $9\frac{309}{500}$

2 a 0.07 b 1.6 c 0.47 d 0.18 e 5.68 f 3.45
g 8.003 h 0.438 i 0.504 j 12.6 k 6.385 l 0.832

3 a $0.\dot{8}$ b 0.625 c 0.714 d $0.\dot{1}\dot{8}$ e $0.91\dot{6}$ f $0.5\dot{3}$

4 a 43% b 85% c 280% d 830% e 87.5% f 116%
g 5% h 387% i 19% j 200.8% k 730% l 20%

5 a $\frac{29}{50}$, 0.58 b $\frac{3}{100}$, 0.03 c $\frac{89}{100}$, 0.89 d $\frac{9}{20}$, 0.45
e $1\frac{3}{4}$, 1.75 f $2\frac{3}{10}$, 2.3 g $\frac{9}{50}$, 0.18 h $1\frac{51}{100}$, 1.51
i $\frac{6}{25}$, 0.24 j $\frac{1}{8}$, 0.125 k $\frac{21}{400}$, 0.0525 l $\frac{11}{40}$, 0.275

6 a $\frac{4}{5}$ b $\frac{3}{5}$ c $\frac{5}{8}$ d $2\frac{1}{4}$ e $1\frac{1}{6}$ f $\frac{1}{5}$
g $1\frac{3}{8}$ h $\frac{1}{7}$ i $\frac{1}{5}$ j $2\frac{2}{3}$ k $\frac{1}{8}$ l $1\frac{3}{5}$

7 a 9:10 b 5:6 c 1:5 d 6:11 e 4:5 f 1:6
g 14:3 h 11:6 i 14:5 j 26:5 k 1:8 l 8:5

8 a $\frac{7}{10}$, 0.67, 55%, 5:11 b 80%, $\frac{3}{4}$, 1:4, 0.075 c $\frac{2}{3}$, 2:4, 0.23, 20%
d 11:4, $2\frac{1}{4}$, 200%, 1.9 e 3:5, 35%, 0.33, $\frac{1}{5}$ f 10:1, 1.0, $\frac{5}{10}$, 10%

Page 60 Assessment

1 a 23.8% b 12.5% c 32.3% d 43.3% e 60.3% f 42.2%

2 a 85% b 18.4% c 8.4% d 20.3% e 20.1% f 51.8%

3 a 50 b 37.4 c 66.7 d 30.45 e 138.88 f 185.4
g 141.9 h 234.57

4 a 72 b 56.1 c 64.8 d 62.25 e 143.52 f 179.76
g 183.15 h 248.5

5 a 77.7% b 50% c 44.8% d 28.7% e 19.3% f 9.8%

6 a 42.3% b 22.7% c 28.7% d 17.9% e 20.9% f 13.4%

7 a K51 b K97.65 c K74.10 d K35.15 e K140.80 f K117.25

8 a K76.50 b K51.80 c K105.90 d K270.60 e K214.10 f K382.40

9 a K1140.91 b 483 members c 1125 litres

Page 61 Assessment

1 b **2** c **3** a **4** b **5** a **6** d
7 b **8** c **9** a **10** a **11** d **12** c
13 c **14** b **15** d **16** d **17** b **18** b

Page 62 Assessment

1 a > b > c < d > e < f < g > h <

2 a 11 b 5 c −2 d −19 e 5 f −7
g 9 h −40 i 15 j 5 k −34 l 35
m 6 n −8 o 31 p −12

3 a 30 b −56 c −72 d 72 e 40 f 150
g 72 h −96 i 4 j −6 k −9 l 4
m 5 n 7 o −8 p 8

4 a $2^2 \times 5$ b $2^4 \times 3$ c $2^2 \times 5^2$ d 3×5^2 e $2^4 \times 5$ f $5^3 \times 2^2$

5 a 82 b 108 c 919 d 101 e 128 f 9
g 968 h 40 i 97 j 375 k 155 l 4095
m 481 n 16 078 o 144 p 6318

6 a 8 b 10 c 9 d 8 e 21 f 30
g 19 h 25 i 40 j 14 k 22 l 21

7 a 6 b 12 c 8 d 80 e 8 f 4
g 4 h 42 i 256 j 1 k 288 l 27

Strand: Space and Shape

Pages 63–64

1 78.5 cm^2 **2** 200.96 cm^2 **3** 28.26 cm^2 **4** 254.34 cm^2
5 706.5 mm^2 **6** 1519.76 mm^2 **7** 1017.36 mm^2 **8** 3215.36 mm^2
9 803.84 m^2 **10** 153.86 m^2 **11** 615.44 m^2 **12** 314 m^2
13 113.04 cm^2 **14** 452.16 cm^2 **15** 2826 cm^2 **16** 1256 cm^2
17 530.66 m^2 **18** 379.94 m^2 **19** 1384.74 m^2 **20** 1962.5 m^2

Page 64

1 25.12 cm^2 **2** 38.47 cm^2 **3** 56.52 cm^2 **4** 84.78 cm^2
5 39.25 cm^2 **6** 50.24 cm^2 **7** 7.07 m^2 **8** 37.68 m^2
9 127.17 m^2 **10** 9.42 m^2 **11** 226.08 m^2 **12** 176.63 m^2
13 981.25 mm^2 **14** 1589.63 mm^2 **15** 1038.57 mm^2 **16** 1017.36 mm^2
17 2119.5 mm^2 **18** 2512 mm^2

Page 65

1 95.52 cm^2 **2** 12.53 m^2 **3** 27.63 m^2 **4** 44.56 cm^2
5 70.65 cm^2 **6** 23.57 m^2 **7** 164.48 m^2 **8** 108.47 m^2
9 87.25 m^2 **10** 357 cm^2 **11** 904.88 cm^2 **12** 69.63 m^2
13 185.12 cm^2 **14** 401.04 cm^2 **15** 157.47 m^2

Page 66

1 125.6 m^2 **2** 122.46 m^2 **3** 103.62 m^2 **4** 93.72 cm^2
5 123.84 cm^2 **6** 821.5 m^2 **7** 310.86 cm^2 **8** 753.6 cm^2
9 226.08 m^2 **10** 523.92 cm^2 **11** 183.5 cm^2 **12** 640.12 m^2
13 843.84 cm^2 **14** 1359.62 m^2 **15** 838.66 m^2

Page 67

1 160 cm^3 **2** 76.5 cm^3 **3** 399 cm^3 **4** 43.75 m^3
5 84 m^3 **6** 312.5 m^3 **7** 93.75 cm^3 **8** 3375 cm^3
9 1296 cm^3 **10** 337.5 m^3 **11** 297 m^3 **12** 316.25 m^3

Page 68

1 1185.75 cm^3 **2** 4950 cm^3 **3** 8960 cm^3 **4** 1365 m^3
5 4725 m^3 **6** 1530 m^3 **7** 5720 cm^3 **8** 864 cm^3
9 577.5 cm^3 **10** 161.5 m^3 **11** 355.25 m^3 **12** 264 m^3

Page 69

1 1177.5 cm^3 **2** 6330.24 m^3 **3** 2260.8 m^3 **4** 3692.64 cm^3
5 2612.48 cm^3 **6** 52 987.5 m^3 **7** 270.04 cm^3 **8** 2089.67 m^3
9 3179.25 cm^3 **10** 427.04 m^3 **11** 395.64 cm^3 **12** 301.44 m^3
13 1295.25 m^3 **14** 12 538.02 cm^3 **15** 155.43 m^3

Page 70

1 1004.8 cm^3 **2** 1519.76 cm^3 **3** 523.33 cm^3 **4** 75.36 m^3
5 150.72 m^3 **6** 2712.96 m^3 **7** 2543.4 cm^3 **8** 2344.53 cm^3
9 3349.33 cm^3 **10** 58.61 m^3 **11** 5626.88 m^3 **12** 640.56 m^3

Page 71

1 40 cm^3 **2** 128 cm^3 **3** 180 cm^3 **4** 99 cm^3
5 85 cm^3 **6** 87 cm^3 **7** 63.33 m^3 **8** 270 m^3
9 320 m^3 **10** 26.67 m^3 **11** 137.5 m^3 **12** 135.33 m^3

Page 72

1 a 30 cm^3 b 865 cm^3 c 7000 cm^3 d 3750 cm^3
e 2 m^3 f 4.5 m^3 g 3 m^3 h 0.9 m^3
i 187 cm^3 j 645 cm^3 k 2390 cm^3 l 324 cm^3
m 8 m^3 n 6.1 m^3 o 40 m^3 p 1983 cm^3
q 518 cm^3 r 0.748 m^3 s 0.5 m^3 t 2.075 m^3
u 15 m^3 v 32.8 m^3 w 0.325 m^3 x 3018 cm^3

2 a 1.327 L b 56 mL c 4000 L (4 kL) d 11 000 L
e 8500 L f 1750 L g 5400 L h 700 L
i 450 mL j 2.716 L k 33 mL l 6.85 L
m 12 900 L n 6570 L o 543 kL p 4180 L
q 917 mL r 1.056 L s 4.387 L t 28 mL
u 115 kL v 1038 kL w 15 300 L x 26 700 L

3 a 415 cm^3 b 10 m^3 c 2.5 m^3 d 78 cm^3
e 27 000 cm^3 f 8.25 m^3 g 16.25 m^3 h 195 000 cm^3
i 880 cm^3 j 11 250 cm^3 k 5640 cm^3 l 36.5 m^3
m 17.75 m^3 n 45.2 m^3 o 0.065 m^3 p 934 cm^3
q 9788 cm^3 r 548 cm^3 s 1.23 m^3 t 490 cm^3
u 6.1 m^3 v 14.3 m^3 w 400 cm^3 x 5800 cm^3

4 a 475 mL b 8500 L c 540 L d 19 mL
e 17 800 L f 5.69 L g 15.45 L h 36 000 L
i 31.732 L j 11 760 L k 1.093 L l 41 300 L
m 600 000 L n 6.46 L o 7 mL p 23 100 L
q 2960 L r 13 090 L s 30 L t 400 L
u 7.2 L v 17.835 L w 22 050 L x 7008 L

Page 73

1 42 litres **2** 75 kilolitres **3** 19.25 kilolitres
4 a 480 000 litres b K36 000
5 a 14 040 cm^3 b 14.04 litres
6 22 608 litres **7** 11.4 litres
8 The dimensions of the rectangular tank must have a product of 10 000 cm^3. Two possible answers are 10 cm × 50 cm × 20 cm and 25 cm × 40 cm × 10 cm.
9 5 metres **10** 843.75 litres
11 a a circular pond with a capacity of 1150 litres
b a rectangular tank with dimensions of 4 m × 5 m × 8 m
c a rectangular tin with dimensions of 12 cm × 8 cm × 9 cm

Page 74

1 a 54 L b 157.5 L c 400.5 L d 288 L e 576 L f 976.5 L
g 679.5 L h 1440 L

2 a 40 gallons b 125.56 gallons c 68.89 gallons d 61.11 gallons
e 100 gallons f 166.22 gallons g 114.67 gallons h 133.78 gallons

3 a 150 gallons b 400 litres c 225 gallons d 2500 litres
e 175 gallons f 1500 litres g 920 litres h 150 gallons
i 75 gallons

4 a 3.6 m b 225 mm c 48 km d 600 mm
e 136 km f 15 m g 208 km h 13.5 m
i 375 mm

5 a 50 miles b 40 inches c 833.33 feet d 166.67 feet
e 26 inches f 162.5 miles g 32.8 inches h 90.63 miles
i 633.33 feet

6 a 16 metres b 30 inches c 20 km d 20 inches
e 90 miles f 35 metres g 300 km h 28 metres
i 55 inches

Pages 75–76

1 a 18.84 cm b 10.99 cm c 17.27 cm d 34.54 cm
e 23.55 cm

2 a 15.7 cm b 56.52 cm c 81.64 cm d 125.6 m
e 53.38 m f 78.5 m g 59.66 cm h 65.94 cm
i 94.2 cm j 87.92 m k 116.18 m l 45.53 m

3 a 18.84 cm b 7.07 cm c 31.4 cm d 11.78 cm
e 25.91 cm f 3.53 cm

Page 77

1 a 13 cm b 59 cm c 26.5 cm d 4.8 cm e 23.2 cm f 15.6 cm
g 76 mm h 212 mm i 47.8 cm j 89.4 cm k 11.7 m l 25.1 m

2 a 18 m b 27 cm c 9.3 m d 3.6 cm
e 40.75 mm f 12.9 m g 7.15 cm h 31.5 m
i 24.55 cm j 67.5 mm k 37.35 m l 4.65 cm

3 a 19.63 cm b 26.38 m c 36.9 cm d 29.2 m
e 14.92 cm f 47.73 m g 80.7 cm h 58.4 m
i 71.59 cm j 100.17 m k 52.44 cm l 127.64 m

4 a 34.23 m b 80.38 cm c 48.67 m d 73.48 cm
e 55.89 m f 66.88 cm g 100.17 m h 60.29 cm
i 30.58 m j 85.16 cm k 16.89 m l 117.69 cm

5 Estimates will vary but the ones here have been made by dividing the circumference by 3.
a 21 cm b 10 cm c 33.3 cm d 16 cm
e 45 cm f 28 cm g 50 m h 39 cm
i 56 m j 70 m k 41.3 m l 83.3 m
m 23.3 cm n 58.3 cm o 66.6 cm p 36.6 cm
q 91.6 cm r 116.6 cm

6 a 11.15 cm b 7.64 cm c 5.1 cm d 27.71 cm
e 20.06 cm f 28.98 cm g 24.52 m h 37.26 m
i 45.22 m j 33.76 m k 39.81 m l 44.27 m
m 80.25 cm n 59.24 cm o 67.2 cm p 54.14 cm
q 66.56 cm r 49.36 cm

7 a 7.96 cm b 5.42 cm c 12.42 cm d 6.69 cm
e 15.13 cm f 10.51 cm g 13.22 m h 17.52 m
i 24.52 m j 20.07 m k 36.95 m l 16.72 m
m 34.72 cm n 28.51 cm o 47.77 cm p 42.36 cm
q 23.57 cm r 51.75 cm

Page 78

1 a 188.4 cm b 94.2 metres

2 23.89 cm **3** 14.13 metres

4 a 235.5 m b 1177.5 m c 8.24 km

5 a 5.5 m b 17.3 m c K77.85

6 a 78.5 m b 91.06 m c K6374.20

7 a 160 cm (1.6 m) b 5 m

8 a 1.68 m b 26.75 cm

Page 79

1 55° **2** 105° **3** 20° **4** 105° **5** 115°
6 105° **7** 65° **8** 100° **9** 40° **10** 65°
11 64° **12** 75° **13** 81° **14** 87° **15** 65°

Page 80

1 a 360° b 540° c 1080° d 900°
e 720° f 1080° g 1260° h 720°

2 a 900° b 540° c 180° d 1080°
e 720° f 360° g 1260° h 540°

Page 81

1 107° **2** 120° **3** 130° **4** 125° **5** 60° **6** 120°
7 150° **8** 105° **9** 130° **10** 100° **11** 100° **12** 120°

Page 82

1 a 130° b 125° c 135° d 135° e 118° f 133°

2 a $x = 100°$ b $x = 150°, y = 105°$ c $x = 140°, y = 40°$
d $x = 80°, y = 100°$ e $x = 108°, y = 72°$ f $x = 95°, y = 66°$

Page 83

1 a 140° b 10° c 60° d 138° e 54° f 74°
g 32° h 51° i 153° j 87° k 62° l 54°

2 a 24° b 53° c 72° d 31° e 69° f 48°
g 36° h 81° i 64° j 18° k 27° l 79°

3 a 96° b 41° c 127° d 71° e 133° f 64°
g 168° h 82° i 109° j 144° k 52° l 37°

Page 84

1 $x = 85°, y = 95°$ **2** $x = 35°, y = 145°$ **3** $x = 110°, y = 70°$
4 $x = 150°, y = 30°$ **5** $x = 80°, y = 100°$ **6** $x = 43°, y = 137°$
7 $x = 45°, y = 40°, z = 95°$ **8** $x = 80°, y = 80°, z = 20°$
9 $x = 66°, y = 44°, z = 70°$ **10** $x = 100°, y = 27°, z = 53°$
11 $w = 80°, x = 18°, y = 60°, z = 22°$ **12** $w = 19°, x = 81°, y = 43°, z = 37°$

Pages 85–86

1 a H b C c E d B e C f G

2 a W b Z c T

3 a A b F c G

4 a $x = 70°$ b $y = 70°$
c $a = 50°, b = 130°$ d $c = 50°, d = 50°$
e $u = 105°, v = 75°, w = 75°$ f $e = 30°, f = 30°$
g $x = 130°, y = 130°, z = 50°$ h $a = 85°, b = 95°, c = 95°$
i $d = 50°, e = 130°, f = 130°$ j $s = 90°, t = 90°, u = 90°$
k $w = 55°, x = 55°$ l $y = 125°, z = 125°$
m $d = 112°, e = 112°, f = 112°$ n $g = 45°, h = 45°, i = 135°$
o $a = 78°, b = 78°, c = 78°$ p $s = 118°, t = 62°, u = 118°$
q $v = 124°, w = 56°$ r $g = 112°, h = 112°$

Page 87

1 $x = 148°, y = 32°, z = 148°$ **2** $x = 63°$
3 $c = 98°, d = 98°$ **4** $a = 28°$ **5** $e = 140°, f = 40°$

6 $w = 75°, x = 27°, y = 78°, z = 78°$ **7** $e = 85°, f = 85°, g = 95°$
8 $d = 13°$ **9** $t = 135°$ **10** $y = 154°$
11 $a = 68°, b = 86°, c = 26°, d = 26°$ **12** $u = 48°, v = 132°, w = 132°$
13 $b = 54°$ **14** $d = 68°, e = 112°$ **15** $n = 41°$
16 $x = 64°, y = 116°, z = 116°$ **17** $t = 78°$ **18** $a = 72°$

Page 88

1 a Yes b No c Yes d Yes e Yes f No
2 a Yes b Yes c No d No e Yes f Yes
3 a No b Yes c Yes d Yes e No f Yes

Page 89

1 a 1:200 000 b 1:250 000 c 1:250 000 d 1:50 000 e 1:75 000 f 1:50 000 g 1:5000 h 1:400 000 i 1:5000 j 1:2500 k 1:4000 l 1:150 000
2 a 1.5 km b 3.75 km c 7.5 km d 0.375 km e 22.5 km f 4.5 km g 11.25 km h 9 km i 1.875 km j 3.375 km k 15 km l 16.5 km
3 a 250 km b 25 km c 400 km d 100 km e 500 km f 175 km g 375 km h 60 km i 625 km j 230 km k 575 km l 470 km
4 a 2 km b 5 km c 1.5 km d 2.5 km e 4 km f 7.5 km g 3.5 km h 5.5 km i 0.25 km j 1.25 km k 3.75 km l 2.25 km
5 a 2 cm b 5 cm c 0.5 cm d 10 cm e 0.25 cm f 2.5 cm g 7.5 cm h 0.75 cm i 1.25 cm j 3.5 cm k 6 cm l 4.75 cm
6 a 5 cm b 2 cm c 0.5 cm d 10 cm e 2.5 cm f 12.5 cm g 6.25 cm h 31.25 cm i 3 cm j 15 cm k 7.5 cm l 3.75 cm
7 a 4 cm b 8 cm c 1 cm d 3 cm e 6 cm f 20 cm g 0.5 cm h 2.5 cm i 3.5 cm j 11 cm k 9 cm l 5.5 cm

Pages 90–92 Assessment

1 a 452.16 m^2 b 1017.36 m^2 c 76.93 cm^2 d 19.63 cm^2 e 44.41 m^2 f 262.72 m^2
2 a 65.94 m^2 b 286 cm^2 c 196.83 cm^2
3 a 184 cm^3 b 3120 cm^3 c 14 130 cm^3 d 640.56 cm^3 e 21 m^3 f 115.5 m^3
4 a 746 cm^3 b 387 cm^3 c 8.6 m^3 d 6.5 m^3 e 4250 cm^3 f 0.85 m^3 g 455 cm^3 h 1765 cm^3
5 a 218 kL b 9000 L c 64 mL d 3.485 L e 17 500 L f 44 mL g 31 200 L h 400 L
6 a 13 cm b 7.4 mm c 22.8 mm d 17.6 cm e 25.2 cm f 59.6 m g 47.3 m h 35.9 m
7 a 10.4 cm b 6.95 cm c 37.3 mm d 57.75 mm e 63.85 m f 53.18 m g 46.92 m h 25.54 cm
8 a 60.92 cm b 40.19 cm c 53.07 cm d 8.16 m e 4.71 m f 4.27 m g 96.59 cm h 83.05 m i 120.39 cm
9 a 28.03 cm, 14.02 cm b 71.66 m, 35.83 m c 61.46 cm, 30.73 cm d 43.63 m, 21.82 m e 65.61 cm, 32.81 cm f 36.31 cm, 18.16 cm g 82.17 cm, 41.09 cm h 20.06 cm, 10.03 cm i 50.64 m, 25.32 m j 74.2 m, 37.1 m k 59.87 m, 29.94 m l 33.12 m, 16.56 m
10 a 12 km b 40 km c 26 km d 3 km e 20 km f 34 km
11 a 2 cm b 8 cm c 5 cm d 3 cm e 0.5 cm f 6 cm
12 a 65° b 55° c 70°
13 a 540° b 1080° c 1440°
14 a 115° b 70° c 150°
15 a $x = 40°, y = 140°$ b $x = 87°, y = 24°, z = 69°$ c $x = 75°, y = 75°, z = 30°$ d $a = 18°$ e $b = 102°$ f $d = 142°$ g $c = 55°, d = 55°$ h $w = 55°, x = 55°$ i $d = 104°, e = 104°, f = 104°$

Strand: Measurement

Pages 93–94

1 a 43.45 kg b 54.4 kg c 38.4 kg d 47.15 kg e 46.31 kg
2 a 2790 g, 4030 g, 4650 g, 6200 g b 2.66 kg, 3.84 kg, 4.43 kg, 5.9 kg c 2.88 kg, 4.16 kg, 4.8 kg, 6.4 kg d 2475 g, 3575 g, 4125 g, 5500 g
3 a 5075.85 kg b 424.15 kg
4 18.4 kg
5 a K163.80 b K58.50 c K93.60 d K210.60 e K128.70 f K280.80
6 K27.50 **7** 4.65 kg
8 a 51.6 tonnes b 22.75 tonnes c 47.25 tonnes d 8.75 tonnes e 22 tonnes f 250.5 tonnes
9 a 4.35 tonnes b 4.75 tonnes c 6.5 tonnes d 1.73 tonnes
10 peanuts: 3.99 tonnes per hectare; taro: 7.25 tonnes per hectare
11 a 378.4 tonnes b 367.4 tonnes c 2.9% decrease
12 a 139.6 kg b 142.8 kg c 15.67 kg d 28.6 kg e 149.92 kg f 39.4 kg g 12.57 kg h 123.06 kg

Page 95

1 a 8 a.m. b 11 a.m. c 2 a.m. d 8 p.m. e 5:30 p.m. f 3:30 a.m. g 3:30 a.m. h 12:15 p.m.
2 a Thursday 9:15 p.m. b Thursday 7:15 a.m. c Thursday 10:15 p.m. d Thursday 2:15 p.m. e Thursday 5:15 a.m. f Thursday 11:15 p.m. g Thursday 8:15 a.m. h Thursday 8:15 p.m.
3 a Sunday 12 noon b Monday 6 a.m. c Monday 8 a.m. d Sunday 10 p.m. e Monday 6 a.m. f Monday 4 a.m. g Sunday 2 p.m. h Sunday 9 p.m.
4 a Monday 10:30 p.m. b Monday 7:30 a.m. c Monday 8:30 p.m. d Tuesday 1:30 a.m. e Monday 3:30 p.m. f Monday 5:30 a.m. g Monday 5:30 a.m. h Monday 11:30 p.m.
5 a Friday 8:30 a.m. b Friday 4:30 a.m. c Thursday 9:30 p.m. d Thursday 3:30 p.m. e Friday 5:30 a.m. f Friday 6:30 a.m. g Thursday 12:30 p.m. h Friday 3:30 a.m.

Page 96

1 a Tuesday b 40.5 hours c K627.75
2 a 0935 b 1345 c 0640 d 1830 e 1205 f 2025 g 1510 h 1955 i 0900 j 1620 k 0750 l 1230

3 a 4:19 p.m. b 5:05 p.m. c 11:25 a.m. d 75 minutes
e 10:02 p.m. f 1:35 p.m.

4 a 1 hour 7 minutes b 1 hour 19 minutes c 38 minutes
d 52 minutes e 19 minutes f 10 minutes
g 56 minutes h 47 minutes i 1 hour 14 minutes
j 31 minutes

5 a Trip A b Trip B c 4 hours 55 minutes
d 3 hours 55 minutes e Trips C, D and F f 1620 to 2035

Page 97

1 1624

2 a 3 hours 20 minutes b 13 hours 20 minutes c 160 hours

3 1731

4 a 5 hours 50 minutes b 17.5 hours c 46 hours 40 minutes

5 11 minutes

6 a K36.75 b K55.13 c K132.30 d K110.25

7 a 8 hours 40 minutes b 1 hour 20 minutes c 40 minutes

8 a 90 minutes, 120 minutes, 150 minutes, 180 minutes, 210 minutes
b 75 minutes, 100 minutes, 125 minutes, 150 minutes, 175 minutes
c 120 minutes, 160 minutes, 200 minutes, 240 minutes, 280 minutes

9 a 138.8 seconds (2 minutes 18.8 seconds)
b 3.7 seconds c 34.7 seconds

Page 98 Assessment

1 a Yes b No c Yes d Yes e No

2 a 9.3 tonnes b 4.7 tonnes c 15.8 tonnes d 9.5 tonnes
e 1.85 tonnes

3 a 1330 b 1900 c 0810 d 0915 e 1540 f 1025
g 1745 h 0705 i 2150 j 1120 k 1355 l 2015

4 a 10:25 a.m. b 6 p.m. c 12:50 p.m. d 3:57 p.m.
e 12:03 p.m. f 1:28 a.m. g 8:54 p.m. h 6:46 a.m.

5 a 7:01 a.m. b 6:22 p.m. c 1:16 p.m. d 5:44 a.m.
e 3:08 p.m. f 7:53 p.m. g 5:29 a.m. h 11:11 a.m.

6 a 2 hours 39 minutes b 4 hours 41 minutes c 3 hours 16 minutes
d 3 hours 29 minutes e 7 hours 23 minutes f 5 hours 47 minutes
g 3 hours 27 minutes h 5 hours 26 minutes i 3 hours 28 minutes

7 a 6 hours 6 minutes b 5 hours 34 minutes

Strand: Chance and Data

Page 99

1 a 9.1, 8, 8, 16 b 15.8, 15.5, 10, 16 c 6.7, 6.5, no mode, 5.3
d 9.1, 9.5, 12.3, 8.2 e 25.7, 25, 25, 15 f 11.1, 11, 16, 11
g 56.7, 55, no mode, 43 h 6.9, 8, 8, 7 i 3.15, 3.25, 4.1, 2.6
j 58, 64, 64, 56 k 56.4, 55, 55, 61 l 83, 96, 96, 87
m 60.3, 68, 43, 77 n 4.7, 3.7, 3.7 & 7.4, 6.1 o 102.2, 106.5, 127, 75

2 a 35 b 45 c 40 d 40

Page 100

1 a

Score	Frequency	Relative frequency
5	3	3 ÷ 35 = 0.09 (9%)
6	5	0.14 (14%)
7	8	0.23 (23%)
8	12	0.34 (34%)
9	5	0.14 (14%)
10	2	0.06 (6%)
Totals	35	100%

b

Score	Frequency	Relative frequency
1–20	2	0.03 (3%)
21–40	3	0.04 (4%)
41–60	15	0.21 (21%)
61–80	27	0.38 (38%)
81–100	25	0.34 (34%)
Totals	72	100%

c

Score	Frequency	Relative frequency
0–5 kg	6	0.1 (10%)
6–10 kg	10	0.16 (16%)
11–15 kg	5	0.08 (8%)
16–20 kg	18	0.3 (30%)
21–25 kg	9	0.15 (15%)
26–30 kg	8	0.13 (13%)
30+ kg	5	0.08 (8%)
Totals	61	100%

d

Score	Frequency	Relative frequency
31°	2	0.07 (7%)
30°	3	0.1 (10%)
29°	7	0.23 (23%)
28°	5	0.17 (17%)
27°	8	0.27 (27%)
26°	4	0.13 (13%)
25°	1	0.03 (3%)
Totals	30	100%

2 a 28% b 22%

Page 101

1 a 500 b 100 c $\frac{84}{100} = \frac{21}{25}$ d 420

2 a 200 b 60 c $\frac{9}{60} = \frac{3}{20}$ d 30

3 a 1200 b 250 c $\frac{180}{250} = \frac{18}{25}$ d 864

4 a 275 b 50 c $\frac{43}{50}$
d 236.5, but because we cannot have $\frac{1}{2}$ of a person we round it to 237.

Page 102

1 a 75°, $\frac{75}{360} = \frac{5}{24}$, K250 b 45°, $\frac{45}{360} = \frac{3}{24}$, K150 c 120°, $\frac{120}{360} = \frac{1}{3}$, K400
d 90°, $\frac{90}{360} = \frac{1}{4}$, K300 e 30°, $\frac{30}{360} = \frac{1}{12}$, K100

2 a 120°, $\frac{120}{360} = \frac{1}{3}$, 20 b 30°, $\frac{30}{360} = \frac{1}{12}$, 5 c 90°, $\frac{90}{360} = \frac{1}{4}$, 15
d 72°, $\frac{72}{360} = \frac{1}{5}$, 12 e 48°, $\frac{48}{360} = \frac{2}{15}$, 8

3 a 132°, $\frac{132}{360} = \frac{11}{30}$, 88 b 96°, $\frac{96}{360} = \frac{4}{15}$, 64 c 75°, $\frac{75}{360} = \frac{5}{24}$, 50
d 42°, $\frac{42}{360} = \frac{7}{60}$, 28 e 15°, $\frac{15}{360} = \frac{1}{24}$, 10

4 a 96°, $\frac{96}{360} = \frac{4}{15}$, 12 b 104°, $\frac{104}{360} = \frac{13}{45}$, 13 c 24°, $\frac{24}{360} = \frac{1}{15}$, 3
d 80°, $\frac{80}{360} = \frac{2}{9}$, 10 e 56°, $\frac{56}{360} = \frac{7}{45}$, 7

Pages 103–104

1 Red: $\frac{30}{100} = \frac{3}{10}$; Blue: $\frac{35}{100} = \frac{7}{20}$; Green: $\frac{15}{100} = \frac{3}{20}$; Yellow: $\frac{20}{100} = \frac{1}{5}$

2 a Number 1 = 11.5%; Number 2 = 13%; Number 3 = 9.5%; Number 4 = 13.5%; Number 5 = 10.5%; Number 6 = 14.5%; Number 7 = 12%; Number 8 = 15.5%
b 56.5% c 52.5%

3 a 60% b 40%

4 Red = 28.3%; Yellow = 36.7%; Blue = 35%

5 a White = 24%; Red = 31.3%; Black = 28.7%; Green = 16%
b 40%

Pages 104–105

1 a $\frac{1}{5}$ b $\frac{1}{3}$ c $\frac{7}{15}$ d $\frac{8}{15}$ e $\frac{4}{5}$ f $\frac{2}{3}$

2 a $\frac{1}{4}$ b $\frac{1}{3}$ c $\frac{5}{12}$ d 0 e 1

3 a $\frac{7}{10}$ b $\frac{1}{2}$ c $\frac{2}{5}$ d $\frac{2}{5}$ e $\frac{9}{10}$ f $\frac{3}{10}$

4 a $\frac{1}{4}$ b $\frac{1}{2}$ c $\frac{1}{4}$ d $\frac{3}{4}$ e $\frac{3}{4}$ f $\frac{1}{4}$

5 a $\frac{1}{200}$ b $\frac{1}{150}$ c $\frac{1}{10}$ d $\frac{3}{50}$ e $\frac{19}{100}$ f $\frac{101}{200}$

6 a $\frac{1}{200}$ b $\frac{1}{50}$ c $\frac{1}{40}$ d $\frac{1}{1000}$ e $\frac{1}{25}$ f $\frac{1}{10}$

7 a $\frac{1}{13}$ b $\frac{1}{4}$ c $\frac{1}{52}$ d $\frac{2}{13}$ e $\frac{1}{26}$ f $\frac{1}{52}$

8 a $\frac{1}{8}$ b $\frac{3}{8}$ c $\frac{1}{8}$ d $\frac{3}{8}$ e $\frac{7}{8}$ f $\frac{1}{2}$

Page 106

1 a 2 500 000 b 12 800 000 c 5 600 000 d 37 700 000
e 4 900 000 f 28 700 000 g 33 400 000 h 8 700 000
i 15 200 000 j 62 200 000 k 6 300 000 l 76 400 000

2 a more b less c less d more e more f more
g less h less i less j more k less l less

3 a less b less c less d less e more f less
g less h more i more j more k less l more

4 a 1 400 000 b 3 000 000 c 4 900 000 d 3 600 000
e 1 200 000 f 5 600 000 g 60 h 225
i 150 j 200 k 100 l 83

5 a 1 627 425 b 2 804 456 c 4 839 131 d 3 724 116
e 1 067 855 f 5 705 756 g 63.41 h 233.26
i 143.56 j 178.63 k 111.18 l 83.66

6 a 51.64 b 23.86 c 76.08 d 33.56 e 92.08 f 47.37
g 19.17 h 67.91 i 38.28 j 29.86 k 56.15 l 85.53
m 16.08 n 64.69 o 31.28 p 70.66

7 a 36.7 b 3.5 c 72.9 d 48.4 e 8.4 f 26.1
g 8.4 h 61.1 i 54.4 j 6.7 k 63.5 l 82.2
m 33.6 n 25.6 o 2.4 p 40.8

Page 107 Assessment

1 a 18.78, 20 b 64.63, 62.5 c 6.55, 6.9 d 8.2, 8.8

2 a 38, 13 b 87 and 71, 34 c 63, 43 d 2.1, 6.6

3 a 2000 b 360 c $\frac{19}{180}$ d 211

4 a 200 b 40 c $\frac{3}{20}$ d 30

5 a Number 1 = 16.1%; Number 2 = 17.8%; Number 3 = 15%; Number 4 = 16.1%; Number 5 = 18.3%; Number 6 = 16.7%
b 49.4% c 33.9%

6 a $\frac{5}{18}$ b $\frac{1}{2}$ c $\frac{2}{9}$ d $\frac{13}{18}$ e $\frac{7}{9}$

7 a $\frac{1}{500}$ b $\frac{2}{125}$ c $\frac{1}{100}$ d $\frac{3}{100}$ e $\frac{1}{250}$ f $\frac{3}{125}$

Strand: Patterns and Algebra

Page 108

1 10, 4, 7, 15, 6, 12

2 21, 31, 27, 45, 39, 57

3 57, 42, 197, 107, 77, 252

4 15, 28, 6, 39, 12, 47

5 30, 41, 38, 25, 13, 31

6 1000, 125, 1728, 8000, 27, 1331

7 60, 35, 100, 20, 40, 15

8 156, 184, 128, 172, 116, 200

9 8, 4, 12, 16, 32, 20

10 88, 67, 79, 1, 37, 52

11 61, 43, 53, 71, 103, 47

12 72, 152, 44, 408, 129, 233

13 3, 2, 0, 13, 5, 9

14 18, 27, 9, 0, 24, 12

15 19, 32, 21, 14, 30, 25

16 106, 56, 126, 86, 246, 606

Page 109

1 $y = 10x$ **2** $b = a + 7$ **3** $g = c - 9$ **4** $s = \frac{d}{7}$

5 $k = h^2$ **6** $p = 3m - 1$ **7** $z = 10w + 2$ **8** $n = f^3$

9 $r = p - 12$ **10** $d = 6c$ **11** $y = \frac{x}{4}$ **12** $h = g^2 + 1$

13 $m = \sqrt{k}$ **14** $t = 5s - 2$ **15** $p = n - 15$ **16** $z = y^2 - 2$

17 $k = 9j$ **18** $d = \frac{b}{8}$

Page 110

1 $f = e + 4$; 61, 40, 82

2 $c = b - 11$; 11, 72, 43

3 $a = s^2$; 100, 900, 144

4 $v = \frac{t}{4}$; 21, 24, 13

5 $y = 12x$; 120, 360, 96

6 $d = 2w + 5$; 27, 65, 39

7 $h = 2g - 1$; 83, 53, 129

8 $m = k + 20$; 59, 74, 82

9 $p = r - 50$; 38, 26, 57

10 $b = \frac{a}{10}$; 21, 9, 12

11 $z = \sqrt[3]{n}$; 5, 3, 6

12 $g = 20f$; 300, 140, 100

13 $d = 3c + 10$; 22, 55, 250

14 $k = 5p - 2$; 28, 58, 198

15 $x = w - 8$; 84, 35, 27

16 $m = f^2 + 1$; 122, 901, 82

17 $z = y^2 - 1$; 224, 80, 35

18 $b = h + 25$; 39, 108, 87

Pages 111–112

1 **a** $4ab$ **b** $7d$ **c** xyz **d** $12p$ **e** $\frac{g}{8}$ **f** $\frac{23}{w}$
g $\frac{ab}{5}$ **h** $\frac{17}{xy}$ **i** $\frac{6z}{3w}$ **j** $\frac{2ab}{c}$ **k** $9cg$ **l** $\frac{m}{4}$
m $\frac{8v}{3}$ **n** $7x + 5y$ **o** $6d$ **p** $\frac{s}{5} - \frac{m}{3}$ **q** $8gh$ **r** $18mn$
s $30ab$ **t** $\frac{d}{5} - 3g$ **u** $\frac{8mn}{s}$ **v** $\frac{15g}{2h}$ **w** $\frac{y}{4} + 3a$ **x** $6k - \frac{12}{z}$

2 **a** $4m^2$ **b** $9d^3$ **c** $5a^2$ **d** $10y^2$ **e** $4a^2$ **f** $5y^3z^2$
g $12h^3$ **h** $8b^4$ **i** $6d^3$ **j** $5a^3b$ **k** $40c^3$ **l** $7w^4$
m $3g^2h^3$ **n** $10y^2d^3$ **o** $21k^3$ **p** $12fgh^2$ **q** $24m^2$ **r** $12w^3$

3 **a** y^8 **b** c^8 **c** k^7 **d** b^5 **e** $6g^5$ **f** $20k^6$
g $18a^3$ **h** $25d^5$ **i** $2m^6$ **j** $4w^3$ **k** $28h^6$ **l** $9p^4$
m $4a^8$ **n** $18b^4$ **o** $6y^6$ **p** $4k^{10}$

4 **a** a^3 **b** d^4 **c** w^2 **d** y^5 **e** $3b^2$ **f** $4k$
g $5h^2$ **h** $3d^4$ **i** $2w^3$ **j** $3g^5$ **k** $4x^2$ **l** $4b^4$
m $4y^8$ **n** d^6 **o** $3w^2$ **p** k

Pages 112–113

1 **a** No **b** Yes **c** Yes **d** No **e** No **f** Yes
g No **h** No **i** Yes **j** Yes **k** Yes **l** No
m Yes **n** Yes **o** Yes **p** No

2 **a** No **b** Yes **c** Yes **d** Yes **e** No **f** No
g No **h** No **i** Yes **j** No **k** Yes **l** Yes
m No **n** Yes **o** Yes **p** Yes

3 **a** $2m$ **b** $4d$ **c** $3ab$ **d** $3h$ **e** $5b$ **f** $3x$
g 0 **h** $7d$ **i** $13k$ **j** $3gh$ **k** $7xw$ **l** $3g$
m $13y$ **n** $7x^2$ **o** $2p$ **p** $6a^3$ **q** $3v$ **r** $3mn$
s $10s$ **t** $4x$ **u** $5d$ **v** $6a^2$ **w** $4gh$ **x** $7k$

4 **a** $7a - b$ **b** $3w + 3x$ **c** $11x + 7y$ **d** $6t + 3c$
e $10 - 3g$ **f** $8h + 3$ **g** $8a + 5b$ **h** $2m + 4n$
i $7y - 5x$ **j** $6k - 3$ **k** $3 + 6w$ **l** $13d + 6$
m $3h + 3s$ **n** $4a + 5b$ **o** $3p + 5pq$ **p** $3ab + 5a$
q $11xy + 8d$ **r** $8gh - 3g$ **s** $2cd + 7c$ **t** $4 + 2y + 3xy$
u $3ab + 5 + 3xy$ **v** $9d + 3gh + 7$ **w** $3b - 2b^2$ **x** $4xy^2 - 5xy$

Pages 113–114

1 **a** $7x + 7y$ **b** $2c + 2d$ **c** $5w + 5z$ **d** $10s + 10t$
e $3g + 3h$ **f** $8p + 8r$ **g** $11n + 11m$ **h** $6k + 6v$
i $5y + 15$ **j** $3x + 21$ **k** $9a + 36$ **l** $4b + 20$
m $cd + 2d$ **n** $ab + 5a$ **o** $gh + 7g$ **p** $wy + 4w$
q $4m + dm$ **r** $35 + 5p$ **s** $5y + yz$ **t** $45 + 9k$
u $6a + 6b$ **v** $4w + 4g$ **w** $8h + 24$ **x** $15 + 3y$

2 **a** $4a - 4b$ **b** $7f - 7g$ **c** $12x - 12y$ **d** $3m - 3n$
e $6p - 6r$ **f** $9c - 9d$ **g** $2w - 2z$ **h** $5s - 5t$
i $8g - 40$ **j** $4h - 24$ **k** $10d - 30$ **l** $6f - 48$
m $by - 2b$ **n** $pr - 7p$ **o** $xy - 12x$ **p** $vw - 3v$
q $5c - cs$ **r** $12 - 3g$ **s** $9c - cd$ **t** $16 - 8m$
u $7g - 7h$ **v** $3n - 3s$ **w** $2a - 16$ **x** $30 - 5b$

3 **a** $20a + 10$ **b** $21b - 35$ **c** $12g + 14$ **d** $40y - 24$
e $18 - 12a$ **f** $18 + 45b$ **g** $40 - 16d$ **h** $18 + 12y$
i $8x + 12z$ **j** $15c - 25d$ **k** $42g + 21h$ **l** $12m - 15n$
m $15p - 3r$ **n** $63a + 7b$ **o** $66m - 11n$ **p** $36y + 9z$
q $30g + 24h$ **r** $21f - 6w$ **s** $24y + 48d$ **t** $70c - 50a$
u $6k - 14v$ **v** $16m + 20p$ **w** $30t - 10s$ **x** $24b + 27d$

4 **a** $4ak + 5bk$ **b** $3ag - 2ah$ **c** $7xy + 4xz$ **d** $5dw - 2dn$
e $3hm - 2gm$ **f** $7dx + 4dy$ **g** $5cg - 4dg$ **h** $8az + 3bz$
i $12dw + 16bw$ **j** $6ak - 10bk$ **k** $20hy + 15gy$ **l** $35dm - 14dk$
m $10xy - 15wx$ **n** $8df + 14cf$ **o** $24mn - 18np$ **p** $8xz + 20xy$
q $18ct + 60dt$ **r** $20gh - 4gk$ **s** $3bd + 21bk$ **t** $30nv - 5mn$

Pages 115–116

1 **a** $x + y$ **b** gh **c** $a - b$ **d** $\frac{w}{7}$
e $d + 2$ **f** $z - 5$ **g** $2k + 7$ **h** $3m - 2$
i $2(s + t)$ **j** $\frac{yz}{2}$ **k** $5g - 3$ **l** $9n + 6$
m $xy - 4$ **n** $\frac{c}{d} + 7$ **o** $p - q$ **p** $\frac{k}{11} + 4$
q $6f$ **r** $\frac{m}{10}$ **s** $g - 8$ **t** $b + 12$

2 **a** $7(b + d)$ **b** $\frac{x + y}{5}$ **c** $\frac{w}{g} + 2$ **d** $p + q - 1$
e $8(a + 5)$ **f** $4(g - 3)$ **g** $mn + 8$ **h** $\frac{b}{4} + 2$
i $\frac{6 - h}{k}$ **j** $t(d + 7)$ **k** $w(y + 3)$ **l** $\frac{cd}{2}$
m $\frac{g + h}{5}$ **n** $\frac{k + 4}{y}$ **o** $2(9 - d)$ **p** $5b + c$

3 **a** subtract 7 from *z* **b** add 8 to *g*
c multiply 4 and *a* **d** divide *b* by 3
e multiply *x* and *y* **f** add *p* to q
g subtract 10 from *k* **h** multiply 5 and *g*, then subtract 2
i add the product of 4 and *d* to 8 **j** add *g* and *h*, then multiply by 6
k subtract *z* from *y*, then multiply by 3 **l** multiply *a*, *b* and *c* together
m multiply *c* and *d*, then add 11 **n** divide the product of *m* and *n* by 2
o divide *t* by 5, then add it to *s* **p** divide *g* by *h*, then add 7
q subtract the product of 6 and *d* from 15
r add 9 to *c*, then multiply by *k*
s subtract 5 from *v*, then multiply by 4
t divide *n* by 3, then subtract it from *f*

4 **a** $a = 6b - 7$ **b** $a = dfg$ **c** $a = \frac{c + 10}{3}$ **d** $a = \frac{x}{y} + 8$
e $a = 4k + d$ **f** $a = b(p - 6)$ **g** $a = 5(y + z)$ **h** $a = \frac{s + t + u}{2}$
i $a = 30 - 4d$ **j** $a = \frac{g}{7} - h$

Pages 116–117

1 **a** $5g + 18$ **b** $8b + 4$ **c** $3x + 27$ **d** $20 + 22p$
e $21 + 31d$ **f** $17y - 4$ **g** $2bw + 7$ **h** $6ad + ag$
i $6wy - 6y$ **j** $m + 8$ **k** $5k + 54$ **l** $19d - 21$
m $3g + 22$ **n** $2 + 5h$ **o** $12k + 16$ **p** $11p - 18$
q $11d - 12$ **r** $8s + 10$ **s** $3f + 18$ **t** $56a$
u $20g$ **v** $15xy + 3y$ **w** $10k + 4pq$ **x** $10cd$

2 a $30g + 49$ b $40k - 17$ c $24h + 35$ d $2y + 19$
e $12x + 64$ f $14 + 10d$ g 54 h $32 + 2g$
i $33 - 43b$ j $8 + 23p$ k $30 + 22k$ l $5a + 6$
m $22 - 6h$ n $24 - 16d$ o 45 p $3xy + 5y$
q $8mn - 8m$ r $21\mathrm{d}k + 9k$ s $22 + 3b$ t $19 + 4y$
u $23z + 8$ v $20c - 6cd$ w $20a - 12ak$ x $22h + 10gh$

3 a $11x^2$ b $2y^2$ c $6b^2$ d $3p^2$
e $6n^2$ f $8a^2$ g $30wx + 15w^2$ h $8d^2 + 6de$
l $8ab + 8a^2$ j $17n + n^2$ k $19h - 15h^2$ l $30g^2$
m $49f^2 + 42f$ n $4k + 4k^2$ o $4m^2 - 5m$ p $5h + 6h^2$
q $7w^2 - 5aw$ r $8ps + 4p^2 + 8$ s $5y + 4g^2$ t $4c + 2c^2$
u $8 + 3a^2$ v $3n^2 + 2f^2$ w $2z^2 + 10t^2$ x $6b^2 + 8p^2$

Pages 117–119

1 a 8 b 2 c 15 d $\frac{3}{5}$ e 27 f 11
g 45 h 40 i 40 j 156 k 16 l 48
m 51 n 5 o 55 p 10

2 a 43 b 42 c 7 d 4 e 45 f 1
g 18 h 21 i 26 j 2 k 39 l 35
m 44 n 106 o 22 p 93

3 a 20 b 5 c 135 d 10 e 275 f 5
g 10 h 95 i 189 j 61 k 60 l 60
m 24 n 12 o 104 p 65

4 a 144 b 78 c 48 d 48 e 91 f 21
g 108 h 24 i 31 j 20 k 29 l 10
m 61 n 42 o 24 p 41

5 a 10 b 50 c 60 d 84 e 120 f 62
g 40 h 6 i 48 j 51 k 36 l 28
m 36 n 50 o 2 p 96

6 a 24 b 128 c 47 d 96 e 10 f 130
g 448 h 37 i 9 j 135 k 8 l 108
m 120 n 46 o 76 p 96

7 a $y = 7, 27, 3, 35, 55$ b $m = 28, 124, 67, 147, 52$
c $a = 7, 12, 3, 20, 9$ d $g = 27, 52, 22, 62, 107$
e $w = 39, 35, 41, 26, 30$ f $d = 8, 16, 28, 0, 12$
g $p = 22, 67, 40, 85, 49$ h $t = 50, 46, 44, 57, 43$
i $f = 10, 25, 15, 30, 40$ j $c = 29, 75, 31, 65, 49$
k $r = 17, 73, 1, 41, 92$ l $y = 50, 65, 35, 100, 80$
m $q = 84, 35, 21, 49, 7$ n $z = 8, 4\frac{1}{2}, 50, 72, 12\frac{1}{2}$
o $d = 38, 66, 42, 126, 94$

Pages 120–21

1 a 8 b 6 c 4 d 5 e 28 f 8
g 3 h 3 i 15 j 24 k 9 l 11

2 a 3 b 4 c 7 d 6 e 9 f 12
g 16 h 8

3 a 705 b 2040 c 1134 d 1950 e 3008 f 1407
g 6345 h 6624

4 a $4(h + 4)$ b $7(w - 4)$ c $3(b + 7)$ d $6(y - 3)$
e $6(3a + 1)$ f $8(4d - 1)$ g $6(5k + 2)$ h $3(3m - 2)$
i $5(3 + 8z)$ j $8(7 - 3f)$ k $9(4 + y)$ l $22(3 - 2s)$

5 a $8(x + y)$ b $3(f - g)$ c $4(3a + 2b)$ d $7(4m - 3n)$
e $5(3w + 7k)$ f $2(13d - 7c)$ g $8(2s + 3t)$ h $4(9p - 7r)$
i $10(9mp)$ j $8(3gh)$ k $2(11bt)$ l $6(4wy)$
m $5(9pq + 5p)$ n $8(4ad - a)$ o $2(7v + 4vw)$ p $16(3s - 2ks)$

Page 122 Assessment

1 a $y = 105, 140, 84, 350, 77, 441$ b $h = 19, 51, 91, 39, 111, 55$

2 a $n = 3m$; 180, 54, 162 b $p = 2k + 1$; 51, 89, 29

3 a $6cd$ b $7w$ c efg d $18y$ e $\frac{b}{10}$ f $\frac{12}{x}$
g $\frac{cd}{7}$ h $\frac{24}{mn}$ i $12t$ j $\frac{5z}{2}$ k $16xy$ l $9fg$
m $5p^2$ n $5c^2$ o $6k^2$ p a^4

4 a $9y + 2w$ b $4d$ c $5m + 4n$ d $9s + 2t$
e $2 + 8g$ f $15 - 5w$ g $4v + 3vw$ h $5a + a^2$

5 a $4c + 4d$ b $7h + 42$ c $3a - 3b$ d $kp - 2k$
e $18g + 12$ f $8d - 10$ g $20m - 8n$ h $18f + 12g$

6 a $3h + 13$ b $10 + 11x$ c $4y + 18$ d $7fg + fh$
e $4wz - 4w$ f $3d^2$ g $10k^2 + 14k$ h $12a^2$

7 a 36 b 216 c 45 d 96 e 126 f 38
g 46 h 18

8 a $5(g + 6)$ b $9(t - 2)$ c $4(w + 6)$ d $7(4s - 1)$
e $5(5 + 7x)$ f $6(6 - 5y)$ g $12(a + \mathrm{b})$ h $7(c - d)$